for

S

A *New York*

Short-listed for the 2019 Royal Society Science Book Prize

"Marvelous . . . An array of witty and astonishing stories . . . Illuminates how calculus has helped bring into being our contemporary world and so many of the instruments whose role we now blithely assume."

—*Washington Post*

"Fascinating anecdotes abound in *Infinite Powers* . . . Strogatz uses the right amount of technical detail to convey complex concepts with clarity . . . Evocatively conveys how calculus illuminates the patterns of the universe, large and small."

—*Nature*

"A brilliant, appealing explanation of how calculus works and why it makes our lives so much better."

—*Saturday Evening Post*

"Strogatz does a great job of explaining a difficult subject . . . He lays out the case that calculus is fundamental to the way we live today . . . A solid choice for readers who want to know what calculus is all about, and for teachers who wish to improve their presentation."

—*Library Journal*

"An energetic effort that successfully communicates the author's love of mathematics."

—*Kirkus Reviews*

"Far-ranging survey . . . Clear and accessible . . . Strogatz successfully illuminates a notoriously complex topic and this work should enhance appreciation for the history behind its innovations."

—*Publishers Weekly*

"Fortunately, we live in an era when a top mathematician can write a book about calculus that is accessible to the mathematically challenged . . . It is the historical detail in the book that not only allows me to follow the math by taking me through how it was discovered, but also sticks in my mind. [Strogatz] makes me want to get out a textbook and start studying calculus."

—*Five Books*

"Wonderful . . . Bringing the insights of calculus—among the most important of all developments in the history of mathematics—to everyone."

—Brian Greene

"The world is a big thing made of an infinite number of infinitely small things. That's the lesson of calculus, the most powerful mathematical technique ever invented. In this engaging book, Steven Strogatz illuminates the importance of calculus and explains its mysteries as only he can."

—Sean Carroll, author of *The Big Picture*

"If calculus is the language of the universe, then Steven Strogatz is its Homer. With verve, insight, and simplicity, he explains the deep ideas underlying one of humankind's greatest intellectual achievements. *Infinite Powers* is an incalculable pleasure."

—Daniel Gilbert, author of *Stumbling on Happiness*

"I've never read a clearer explanation of calculus or the significant powers we gained by harnessing infinity. Michael and I highly recommend it!"

—Vsauce

"Are you one of those people who always said you'd someday learn calculus? Well, someday is here, thanks to Steven Strogatz's wide-ranging, humane, thoroughly readable take on one of the greatest ideas our species has ever produced."

—Jordan Ellenberg, author of *How Not to Be Wrong*

"This is a glorious book. Steven Strogatz manages to unmask the true hidden wonder and delightful simplicity of calculus. *Infinite Powers* is a master class in accessible math writing and a perfect read for anyone who feels like they never quite understood what all the fuss was about. It had me leaping for joy."

—Hannah Fry, author of *Hello World*

"Warning: this book is dangerous. It will make you love mathematics. Even more, there is a nonzero risk it will turn you into a mathematician."

—Nassim Nicholas Taleb, author of *The Black Swan*

"In this tour de force, Steve Strogatz shares his love as well as his deep understanding of calculus and mathematics more generally. An elegant and ebullient book, *Infinite Powers* speaks to everyone, reminding us why mathematics matters in a practical sense, while all the time highlighting the cleverness and especially the beauty involved."

—Lisa Randall, Frank B. Baird, Jr., Professor of Physics, Harvard University, author of *Warped Passages* and *Dark Matter and the Dinosaurs*

"This could be the most fascinating book I have ever read. If you have even the slightest curiosity about math and its role in this world, I implore you to read this amazing book. Every teacher, every student, and every citizen will be better for it."

—Jo Boaler, author of *Mathematical Mindsets,* professor of mathematics education, Stanford University, and cofounder of youcubed.org

"Steven Strogatz is a world-class mathematician and a world-class science writer. With a light touch and razor-sharp clarity, he brilliantly filters his deep knowledge of calculus into an engaging epic that tells the remarkable story of a mathematical breakthrough that changed the world—and continues to do so."

—Alex Bellos, author of *Here's Looking at Euclid* and *The Grapes of Math*

"Reading *Infinite Powers,* I was reminded why Steve Strogatz is, at present, the best mathematician among writers and the best writer among mathematicians."

—William Dunham, Bryn Mawr College, author of *Journey Through Genius* and *The Calculus Gallery*

"*Infinite Powers* is simple, lucid, amusing, informative, and a pleasure to read. If you want to know where calculus came from, how it works, what it's good for, and where it's going next, this is the book for you."

—Ian Stewart, professor emeritus of mathematics, University of Warwick, author of *Significant Figures*

"A highly readable account of calculus and its modern applications—all done with the human touch."

—Dr. David Acheson, author of *The Calculus Story*

Infinite Powers

Infinite Powers

How Calculus Reveals the Secrets of the Universe

STEVEN STROGATZ

An Imprint of HarperCollins*Publishers*
Boston New York

First Mariner edition 2020

For information, address HarperCollins Publishers,
195 Broadway, New York, NY 10007.

Mariner
An Imprint of HarperCollins Publishers, registered in the United States of America and/or other jurisdictions.

www.marinerbooks.com

Library of Congress Cataloging-in-Publication Data
Names: Strogatz, Steven H. (Steven Henry), author.
Title: Infinite powers : how calculus reveals the secrets of the universe / Steven Strogatz.
Description: Boston : HarperCollins Publishers, 2019. |
Includes bibliographical references and index.
Identifiers: LCCN 2018042561 (print) | LCCN 2018049721 (ebook) |
ISBN 9781328880017 (ebook) | ISBN 9781328879981 (hardcover) |
ISBN 9780358299288 (pbk.)
Subjects: LCSH: Calculus. | Calculus—History. | Archimedes. | Differential calculus.
Classification: LCC QA303.2 (ebook) | LCC QA303.2 .S78 2010 (print) |
DDC 515—dc23
LC record available at https://lccn.loc.gov/2018042561

Book design by Christopher Granniss

Printed in the United States of America

23 24 25 26 27 LBC 10 9 8 7 6

Contents

Introduction

Without calculus, we wouldn't have cell phones, computers, or microwave ovens. We wouldn't have radio. Or television. Or ultrasound for expectant mothers, or GPS for lost travelers. We wouldn't have split the atom, unraveled the human genome, or put astronauts on the moon. We might not even have the Declaration of Independence.

It's a curiosity of history that the world was changed forever by an arcane branch of mathematics. How could it be that a theory originally about shapes ultimately reshaped civilization?

The essence of the answer lies in a quip that the physicist Richard Feynman made to the novelist Herman Wouk when they were discussing the Manhattan Project. Wouk was doing research for a big novel he hoped to write about World War II, and he went to Caltech to interview physicists who had worked on the bomb, one of whom was Feynman. After the interview, as they were parting, Feynman asked Wouk if he knew calculus. No, Wouk admitted, he didn't. "You had better learn it," said Feynman. "It's the language God talks."

For reasons nobody understands, the universe is deeply mathematical. Maybe God made it that way. Or maybe it's the only way a universe with us in it could be, because nonmathematical universes can't harbor life intelligent enough to ask the question. In any case, it's a mysterious and marvelous fact that our universe obeys laws of nature that always turn out to be expressible in the language of calculus as sentences called differential equations. Such equations describe the difference between something right now and the same thing an instant later or between something right here and the same thing infinitesimally close by. The details differ depending on what part of nature we're talking about, but the structure of the laws is always the same. To put this awesome assertion another way, there seems to be something like a code to the universe, an operating system that animates everything from moment to moment and place to place. Calculus taps into this order and expresses it.

Isaac Newton was the first to glimpse this secret of the universe. He found that the orbits of the planets, the rhythm of the tides, and the trajectories of cannonballs could all be described, explained, and predicted by a small set of differential equations. Today we call them Newton's laws of motion and gravity. Ever since Newton, we have found that the same pattern holds whenever we uncover a new part of the universe. From the old elements of earth, air, fire, and water to the latest in electrons, quarks, black holes, and superstrings, every inanimate thing in the universe bends to the rule of differential equations. I bet this is what Feynman meant when he said that calculus is the language God talks. If anything deserves to be called the secret of the universe, calculus is it.

By inadvertently discovering this strange language, first in a corner of geometry and later in the code of the universe, then by learning to speak it fluently and decipher its idioms and nuances, and finally by harnessing its forecasting powers, humans have used calculus to remake the world.

That's the central argument of this book.

If it's right, it means the answer to the ultimate question of life, the universe, and everything is not 42, with apologies to fans of Douglas Adams and *The Hitchhiker's Guide to the Galaxy*. But Deep Thought was on the right track: the secret of the universe is indeed mathematical.

Calculus for Everyone

Feynman's quip about God's language raises many profound questions. What is calculus? How did humans figure out that God speaks it (or, if you prefer, that the universe runs on it)? What are differential equations and what have they done for the world, not just in Newton's time but in our own? Finally, how can any of these stories and ideas be conveyed enjoyably and intelligibly to readers of goodwill like Herman Wouk, a very thoughtful, curious, knowledgeable person with little background in advanced math?

In a coda to the story of his encounter with Feynman, Wouk wrote that he didn't get around to even trying to learn calculus for fourteen years. His big novel ballooned into two big novels — *Winds of War* and *War and Remembrance*, each about a thousand pages. Once those were finally done, he tried to teach himself by reading books with titles like *Calculus Made Easy*—but no luck there. He poked around in a few textbooks, hoping, as he put it, "to come across one that might help a mathematical ignoramus like me, who had spent his college years in the humanities—i.e., literature and philosophy—in an adolescent quest for the meaning of existence, little knowing that calculus, which I had heard of as a difficult bore leading nowhere, was the language God talks." After the textbooks proved impenetrable, he hired an Israeli math tutor, hoping to pick up a little calculus and improve his spoken Hebrew on the side, but both hopes ran aground. Finally, in desperation, he audited a high-school calculus class, but he fell too far behind and had to give up after a couple of months. The kids clapped for him on his way out. He said it was like sympathy applause for a pitiful showbiz act.

I've written *Infinite Powers* in an attempt to make the greatest ideas and stories of calculus accessible to everyone. It shouldn't be necessary to endure what Herman Wouk did to learn about this landmark in human history. Calculus is one of humankind's most inspiring collective achievements. It isn't necessary to learn how to do calculus to appreciate it, just as it isn't necessary to learn how to prepare fine cuisine to enjoy eating it. I'm going to try to explain everything we'll need with the help of pictures, metaphors, and anecdotes. I'll also walk us through some of the finest equations and proofs ever created, because how could we visit a gallery without seeing its masterpieces? As for Herman Wouk, he is 103 years old as of this writing. I don't know if he's learned calculus yet, but if not, this one's for you, Mr. Wouk.

The World According to Calculus

As should be obvious by now, I'll be giving an applied mathematician's take on the story and significance of calculus. A historian of mathematics would tell it differently. So would a pure mathematician. What fascinates me as an applied mathematician is the push and pull between the real world around us and the ideal world in our heads. Phenomena out there guide the mathematical questions we ask; conversely, the math we imagine sometimes foreshadows what actually happens out there in reality. When it does, the effect is uncanny.

To be an applied mathematician is to be outward-looking and intellectually promiscuous. To those in my field, math is not a pristine, hermetically sealed world of theorems and proofs echoing back on themselves. We embrace all kinds of subjects: philosophy, politics, science, history, medicine, all of it. That's the story I want to tell — the world according to calculus.

This is a much broader view of calculus than usual. It encompasses the many cousins and spinoffs of calculus, both within mathematics and in the adjacent disciplines. Since this big-tent view is unconventional, I want to make sure it doesn't cause any confusion. For example, when I said earlier that without calculus we wouldn't have computers and cell phones and so on, I certainly didn't mean to suggest that calculus produced all these wonders by itself. Far from it. Science and technology were essential partners — and arguably the stars of the show. My point is merely that calculus has also played a crucial role, albeit often a supporting one, in giving us the world we know today.

Take the story of wireless communication. It began with the discovery of the laws of electricity and magnetism by scientists like Michael Faraday and André-Marie Ampère. Without their observations and tinkering, the crucial facts about magnets, electrical currents, and their invisible force fields would have remained unknown, and the possibility of wireless communication would never have been realized. So, obviously, experimental physics was indispensable here.

But so was calculus. In the 1860s, a Scottish mathematical physicist named James Clerk Maxwell recast the experimental laws of electricity and magnetism into a symbolic form that could be fed into the maw of calculus. After some churning, the maw disgorged an equation that didn't make sense. Apparently something was missing in the physics. Maxwell suspected that Ampère's law was the culprit. He tried patching it up by including a new term in his equation—a hypothetical current that would resolve the contradiction—and then let calculus churn again. This time it spat out a sensible result, a simple, elegant wave equation much like the equation that describes the spread of ripples on a pond. Except Maxwell's result was predicting a new kind of wave, with electric and magnetic fields dancing together in a pas de deux. A changing electric field would generate a changing magnetic field, which in turn would regenerate the electric field, and so on, each field bootstrapping the other forward, propagating together as a wave of traveling energy. And when Maxwell calculated the speed of this wave, he found—in what must have been one of the greatest Aha! moments in history—that it moved at the speed of light. So he used calculus not only to predict the existence of electromagnetic waves but also to solve an age-old mystery: What was the nature of light? Light, he realized, was an electromagnetic wave.

Maxwell's prediction of electromagnetic waves prompted an experiment by Heinrich Hertz in 1887 that proved their existence. A decade later, Nikola Tesla built the first radio communication system, and five years after that, Guglielmo Marconi transmitted the first wireless messages across the Atlantic. Soon came television, cell phones, and all the rest.

Clearly, calculus could not have done this alone. But equally clearly, none of it would have happened *without* calculus. Or, perhaps more accurately, it might have happened, but only much later, if at all.

Calculus Is More than a Language

The story of Maxwell illustrates a theme we'll be seeing again and again. It's often said that mathematics is the language of science. There's a great deal of truth to that. In the case of electromagnetic waves, it was a key first step for Maxwell to translate the laws that had been discovered experimentally into equations phrased in the language of calculus.

But the language analogy is incomplete. Calculus, like other forms of mathematics, is much more than a language; it's also an incredibly powerful system of reasoning. It lets us transform one equation into another by performing various symbolic operations on them, operations subject to certain rules. Those rules are deeply rooted in logic, so even though it might seem like we're just shuffling symbols around, we're actually constructing long chains of logical inference. The symbol shuffling is useful shorthand, a convenient way to build arguments too intricate to hold in our heads.

If we're lucky and skillful enough—if we transform the equations in just the right way—we can get them to reveal their hidden implications. To a mathematician, the process feels almost palpable. It's as if we're manipulating the equations, massaging them, trying to relax them enough so that they'll spill their secrets. We want them to open up and talk to us.

Creativity is required, because it often isn't clear which manipulations to perform. In Maxwell's case, there were countless ways to transform his equations, all of which would have been logically acceptable but only some of which would have been scientifically revealing. Given that he didn't even know what he was searching for, he might easily have gotten nothing out of his equations but incoherent mumblings (or the symbolic equivalent thereof). Fortunately, however, they did have a secret to reveal. With just the right prodding, they gave up the wave equation.

At that point the linguistic function of calculus took over again. When Maxwell translated his abstract symbols back into reality, they predicted that electricity and magnetism could propagate together as a wave of invisible energy moving at the speed of light. In a matter of decades, this revelation would change the world.

Unreasonably Effective

It's eerie that calculus can mimic nature so well, given how different the two domains are. Calculus is an imaginary realm of symbols and logic; nature is an actual realm of forces and phenomena. Yet somehow, if the translation from reality into symbols is done artfully enough, the logic of calculus can use one real-world truth to generate another. Truth in, truth out. Start with something that is empirically true and symbolically formulated (as Maxwell did with the laws of electricity and magnetism), apply the right logical manipulations, and out comes another empirical truth, possibly a new one, a fact about the universe that nobody knew before (like the existence of electromagnetic waves). In this way, calculus lets us peer into the future and predict the unknown. That's what makes it such a powerful tool for science and technology.

But why should the universe respect the workings of any kind of logic, let alone the kind of logic that we puny humans can muster? This is what Einstein marveled at when he wrote, "The eternal mystery of the world is its comprehensibility." And it's what Eugene Wigner meant in his essay "On the Unreasonable Effectiveness of Mathematics in the Natural Sciences" when he wrote, "The miracle of the appropriateness of the language of mathematics for the formulation of the laws of physics is a wonderful gift which we neither understand nor deserve."

This sense of awe goes way back in the history of mathematics. According to legend, Pythagoras felt it around 550 BCE when he and his disciples discovered that music was governed by the ratios of whole numbers. For instance, imagine plucking a guitar string. As the string vibrates, it emits a certain note. Now put your finger on a fret exactly halfway up the string and pluck it again. The vibrating part of the string is now half as long as it used to be—a ratio of 1 to 2—and it sounds precisely an octave higher than the original note (the musical distance from one *do* to the next in the *do-re-mi-fa-sol-la-ti-do* scale). If instead the vibrating string is ⅔ of its original length, the note it makes goes up by a fifth (the interval from *do* to *sol;* think of the first two notes of the Star Wars theme). And if the vibrating part is ¾ as long as it was before, the note goes up by a fourth (the interval between the first two notes of "Here Comes the Bride"). The ancient Greek musicians knew about the melodic concepts of octaves, fourths, and fifths and considered them beautiful. This unexpected link between music (the harmony of this world) and numbers (the harmony of an imagined world) led the Pythagoreans to the mystical belief that *all* is number. They are said to have believed that even the planets in their orbits made music, the music of the spheres.

Ever since then, many of history's greatest mathematicians and scientists have come down with cases of Pythagorean fever. The astronomer Johannes Kepler had it bad. So did the physicist Paul Dirac. As we'll see, it drove them to seek, and to dream, and to long for the harmonies of the universe. In the end it pushed them to make their own discoveries that changed the world.

The Infinity Principle

To help you understand where we're headed, let me say a few words about what calculus is, what it wants (metaphorically speaking), and what distinguishes it from the rest of mathematics. Fortunately, a single big, beautiful idea runs through the subject from beginning to end. Once we become aware of this idea, the structure of calculus falls into place as variations on a unifying theme.

Alas, most calculus courses bury the theme under an avalanche of formulas, procedures, and computational tricks. Come to think of it, I've never seen it spelled out anywhere even though it's part of calculus culture and every expert knows it implicitly. Let's call it the Infinity Principle. It will guide us on our journey just as it guided the development of calculus itself, conceptually as well as historically. I'm tempted to state it right now, but at this point it would sound like mumbo jumbo. It will be easier to appreciate if we inch our way up to it by asking what calculus wants . . . and how it gets what it wants.

In a nutshell, calculus wants to make hard problems simpler. It is utterly obsessed with simplicity. That might come as a surprise to you, given that calculus has a reputation for being complicated. And there's no denying that some of its leading textbooks exceed a thousand pages and weigh as much as bricks. But let's not be judgmental. Calculus can't help how it looks. Its bulkiness is unavoidable. It looks complicated because it's trying to tackle complicated problems. In fact, it has tackled and solved some of the most difficult and important problems our species has ever faced.

Calculus succeeds by breaking complicated problems down into simpler parts. That strategy, of course, is not unique to calculus. All good problem-solvers know that hard problems become easier when they're split into chunks. The truly radical and distinctive move of calculus is that it takes this divide-and-conquer strategy to its utmost extreme—*all the way out to infinity*. Instead of cutting a big problem into a handful of bite-size pieces, it keeps cutting and cutting relentlessly until the problem has been chopped and pulverized into its tiniest conceivable parts, leaving infinitely many of them. Once that's done, it solves the original problem for all the tiny parts, which is usually a much easier task than solving the initial giant problem. The remaining challenge at that point is to put all the tiny answers back together again. That tends to be a much harder step, but at least it's not as difficult as the original problem was.

Thus, calculus proceeds in two phases: cutting and rebuilding. In mathematical terms, the cutting process always involves infinitely fine subtraction, which is used to quantify the differences between the parts. Accordingly, this half of the subject is called *differential* calculus. The reassembly process always involves infinite addition, which integrates the parts back into the original whole. This half of the subject is called *integral* calculus.

This strategy can be used on anything that we can imagine slicing endlessly. Such infinitely divisible things are called *continua* and are said to be *continuous*, from the Latin roots *con* (together with) and *tenere* (hold), meaning uninterrupted or holding together. Think of the rim of a perfect circle, a steel girder in a suspension bridge, a bowl of soup cooling off on the kitchen table, the parabolic trajectory of a javelin in flight, or the length of time you have been alive. A shape, an object, a liquid, a motion, a time interval—all of them are grist for the calculus mill. They're all continuous, or nearly so.

Notice the act of creative fantasy here. Soup and steel are not really continuous. At the scale of everyday life, they appear to be, but at the scale of atoms or superstrings, they're not. Calculus ignores the inconvenience posed by atoms and other uncuttable entities, not because they don't exist but because it's useful to pretend that they don't. As we'll see, calculus has a penchant for useful fictions.

More generally, the kinds of entities modeled as continua by calculus include almost anything one can think of. Calculus has been used to describe how a ball rolls continuously down a ramp, how a sunbeam travels continuously through water, how the continuous flow of air around a wing keeps a hummingbird or an airplane aloft, and how the concentration of HIV virus particles in a patient's bloodstream plummets continuously in the days after he or she starts combination-drug therapy. In every case the strategy remains the same: split a complicated but continuous problem into infinitely many simpler pieces, then solve them separately and put them back together.

Now we're finally ready to state the big idea.

The Infinity Principle

To shed light on any continuous shape, object, motion, process, or phenomenon—no matter how wild and complicated it may appear—reimagine it as an infinite series of simpler parts, analyze those, and then add the results back together to make sense of the original whole.

The Golem of Infinity

The rub in all of this is the need to cope with infinity. That's easier said than done. Although the carefully controlled use of infinity is the secret to calculus and the source of its enormous predictive power, it is also calculus's biggest headache. Like Frankenstein's monster or the golem in Jewish folklore, infinity tends to slip out of its master's control. As in any tale of hubris, the monster inevitably turns on its maker.

The creators of calculus were aware of the danger but still found infinity irresistible. Sure, occasionally it ran amok, leaving paradox, confusion, and philosophical havoc in its wake. Yet after each of these episodes, mathematicians always managed to subdue the monster, rationalize its behavior, and put it back to work. In the end, everything always turned out fine. Calculus gave the right answers, even when its creators couldn't explain why. The desire to harness infinity and exploit its power is a narrative thread that runs through the whole twenty-five-hundred-year story of calculus.

All this talk of desire and confusion might seem out of place, given that mathematics is usually portrayed as exact and impeccably rational. It is rational, but not always initially. Creation is intuitive; reason comes later. In the story of calculus, more than in other parts of mathematics, logic has always lagged behind intuition. This makes the subject feel especially human and approachable, and its geniuses more like the rest of us.

Curves, Motion, and Change

The Infinity Principle organizes the story of calculus around a methodological theme. But calculus is as much about mysteries as it is about methodology. Three mysteries above all have spurred its development: the mystery of curves, the mystery of motion, and the mystery of change.

The fruitfulness of these mysteries has been a testament to the value of pure curiosity. Puzzles about curves, motion, and change might seem unimportant at first glance, maybe even hopelessly esoteric. But because they touch on such rich conceptual issues and because mathematics is so deeply woven into the fabric of the universe, the solution to these mysteries has had far-reaching impacts on the course of civilization and on our everyday lives. As we'll see in the chapters ahead, we reap the benefits of these investigations whenever we listen to music on our phones, breeze through the line at the supermarket thanks to a laser checkout scanner, or find our way home with a GPS gadget.

It all started with the mystery of curves. Here I'm using the term *curves* in a very loose sense to mean any sort of curved line, curved surface, or curved solid—think of a rubber band, a wedding ring, a floating bubble, the contours of a vase, or a solid tube of salami. To keep things as simple as possible, the early geometers typically concentrated on abstract, idealized versions of curved shapes and ignored thickness, roughness, and texture. The surface of a mathematical sphere, for instance, was imagined to be an infinitesimally thin, smooth, perfectly round membrane with none of the thickness, bumpiness, or hairiness of a coconut shell. Even under these idealized assumptions, curved shapes posed baffling conceptual difficulties because they weren't made of straight pieces. Triangles and squares were easy. So were cubes. They were composed of straight lines and flat pieces of planes joined together at a small number of corners. It wasn't hard to figure out their perimeters or surface areas or volumes. Geometers all over the world—in ancient Babylon and Egypt, China and India, Greece and Japan—knew how to solve problems like these. But round things were brutal. No one could figure out how much surface area a sphere had or how much volume it could hold. Even finding the circumference and area of a circle was an insurmountable problem in the old days. There was no way to get started. There were no straight pieces to latch onto. Anything that was curved was inscrutable.

So this is how calculus began. It grew out of geometers' curiosity and frustration with roundness. Circles and spheres and other curved shapes were the Himalayas of their era. It wasn't that they posed important practical issues, at least not at first. It was simply a matter of the human spirit's thirst for adventure. Like explorers climbing Mount Everest, geometers wanted to solve curves because they were there.

The breakthrough came from insisting that curves *were* actually made of straight pieces. It wasn't true, but one could pretend that it was. The only hitch was that those pieces would then have to be infinitesimally small and infinitely numerous. Through this fantastic conception, integral calculus was born. This was the earliest use of the Infinity Principle. The story of how it developed will occupy us for several chapters, but its essence is already there, in embryonic form, in a simple, intuitive insight: If we zoom in closely enough on a circle (or anything else that is curved and smooth), the portion of it under the microscope begins to look straight and flat. So in principle, at least, it should be possible to calculate whatever we want about a curved shape by adding up all the straight little pieces. Figuring out exactly how to do this — no easy feat — took the efforts of the world's greatest mathematicians over many centuries. Collectively, however, and sometimes through bitter rivalries, they eventually began to make headway on the riddle of curves. Spinoffs today, as we'll see in chapter 2, include the math needed to draw realistic-looking hair, clothing, and faces of characters in computer-animated movies and the calculations required for doctors to perform facial surgery on a virtual patient before they operate on the real one.

The quest to solve the mystery of curves reached a fever pitch when it became clear that curves were much more than geometric diversions. They were a key to unlocking the secrets of nature. They arose naturally in the parabolic arc of a ball in flight, in the elliptical orbit of Mars as it moved around the sun, and in the convex shape of a lens that could bend and focus light where it was needed, as was required for the burgeoning development of microscopes and telescopes in late Renaissance Europe.

And so began the second great obsession: a fascination with the mysteries of motion on Earth and in the solar system. Through observation and ingenious experiments, scientists discovered tantalizing numerical patterns in the simplest moving things. They measured the swinging of a pendulum, clocked the accelerating descent of a ball rolling down a ramp, and charted the stately procession of planets across the sky. The patterns they found enraptured them — indeed, Johannes Kepler fell into a state of self-described "sacred frenzy" when he found his laws of planetary motion — because those patterns seemed to be signs of God's handiwork. From a more secular perspective, the patterns reinforced the claim that nature was deeply mathematical, just as the Pythagoreans had maintained. The only catch was that nobody could explain the marvelous new patterns, at least not with the existing forms of math. Arithmetic and geometry were not up to the task, even in the hands of the greatest mathematicians.

The trouble was that the motions weren't steady. A ball rolling down a ramp kept changing its speed, and a planet revolving around the sun kept changing its direction of travel. Worse yet, the planets moved faster when they got close to the sun and slowed down as they receded from it. There was no known way to deal with motion that kept changing in ever-changing ways. Earlier mathematicians had worked out the mathematics of the most trivial kind of motion, namely, motion at a constant speed where distance equals rate times time. But when speed changed and kept on changing continuously, all bets were off. Motion was proving to be as much of a conceptual Mount Everest as curves were.

As we'll see in the middle chapters of this book, the next great advances in calculus grew out of the quest to solve the mystery of motion. The Infinity Principle came to the rescue, just as it had for curves. This time the act of wishful fantasy was to pretend that motion at a changing speed was made up of infinitely many, infinitesimally brief motions at a *constant* speed. To visualize what this would mean, imagine being in a car with a jerky driver at the wheel. As you anxiously watch the speedometer, it moves up and down with every jerk. But over a millisecond, even the jerkiest driver can't make the speedometer needle move by much. And over an interval much shorter than that—an infinitesimal time interval—the needle won't move at all. Nobody can tap the gas pedal that fast.

These ideas coalesced in the younger half of calculus, differential calculus. It was precisely what was needed to work with the infinitesimally small changes of time and distance that arose in the study of ever-changing motion as well as with the infinitesimal straight pieces of curves that arose in analytic geometry, the newfangled study of curves defined by algebraic equations that was all the rage in the first half of the 1600s. Yes, at one time, algebra was a craze, as we'll see.

Its popularity was a boon for all fields of mathematics, including geometry, but it also created an unruly jungle of new curves to explore. Thus, the mysteries of curves and motion collided. They were now both at the center stage of calculus in the mid-1600s, banging into each other, creating mathematical mayhem and confusion. Out of the tumult, differential calculus began to flower, but not without controversy. Some mathematicians were criticized for playing fast and loose with infinity. Others derided algebra as a scab of symbols. With all the bickering, progress was fitful and slow.

And then a child was born on Christmas Day. This young messiah of calculus was an unlikely hero. Born premature and fatherless and abandoned by his mother at age three, he was a lonesome boy with dark thoughts who grew into a secretive, suspicious young man. Yet Isaac Newton would make a mark on the world like no one before or since.

First, he solved the holy grail of calculus: he discovered how to put the pieces of a curve back together again—and how to do it easily, quickly, and systematically. By combining the symbols of algebra with the power of infinity, he found a way to represent any curve as a sum of infinitely many simpler curves described by powers of a variable x, like x^2, x^3, x^4, and so on. With these ingredients alone, he could cook up any curve he wanted by putting in a pinch of x and a dash of x^2 and a heaping tablespoon of x^3. It was like a master recipe and a universal spice rack, butcher shop, and vegetable garden, all rolled into one. With it he could solve any problem about shapes or motions that had ever been considered.

Then he cracked the code of the universe. Newton discovered that motion of any kind always unfolds one infinitesimal step at a time, steered from moment to moment by mathematical laws written in the language of calculus. With just a handful of differential equations (his laws of motion and gravity), he could explain everything from the arc of a cannonball to the orbits of the planets. His astonishing "system of the world" unified heaven and earth, launched the Enlightenment, and changed Western culture. Its impact on the philosophers and poets of Europe was immense. He even influenced Thomas Jefferson and the writing of the Declaration of

Independence, as we'll see. In our own time, Newton's ideas underpinned the space program by providing the mathematics necessary for trajectory design, the work done at NASA by African-American mathematician Katherine Johnson and her colleagues (the heroines of the book and hit movie *Hidden Figures*).

With the mysteries of curves and motion now settled, calculus moved on to its third lifelong obsession: the mystery of change. It's a cliché, but it's true all the same — nothing is constant but change. It's rainy one day and sunny the next. The stock market rises and falls. Emboldened by the Newtonian paradigm, the later practitioners of calculus asked: Are there laws of change similar to Newton's laws of motion? Are there laws for population growth, the spread of epidemics, and the flow of blood in an artery? Can calculus be used to describe how electrical signals propagate along nerves or to predict the flow of traffic on a highway?

By pursuing this ambitious agenda, always in cooperation with other parts of science and technology, calculus has helped make the world modern. Using observation and experiment, scientists worked out the laws of change and then used calculus to solve them and make predictions. For example, in 1917 Albert Einstein applied calculus to a simple model of atomic transitions to predict a remarkable effect called stimulated emission (which is what the *s* and *e* stand for in *laser*, an acronym for *light amplification by stimulated emission of radiation*). He theorized that under certain circumstances, light passing through matter could stimulate the production of more light at the same wavelength and moving in the same direction, creating a cascade of light through a kind of chain reaction that would result in an intense, coherent beam. A few decades later, the prediction proved to be accurate. The first working lasers were built in the early 1960s. Since then, they have been used in everything from compact-disc players and laser-guided weaponry to supermarket bar-code scanners and medical lasers.

The laws of change in medicine are not as well understood as those in physics. Yet even when applied to rudimentary models, calculus has been able to make lifesaving contributions. For example, in chapter 8 we'll see how a differential-equation model developed by an immunologist and an AIDS researcher played a part in shaping the modern three-drug combination therapy for patients infected with HIV. The insights provided by the model overturned the prevailing view that the virus was lying dormant in the body; in fact, it was in a raging battle with the immune system every minute of every day. With the new understanding that calculus helped provide, HIV infection has been transformed from a near-certain death sentence to a manageable chronic disease—at least for those with access to combination-drug therapy.

Admittedly, some aspects of our ever-changing world lie beyond the approximations and wishful thinking inherent in the Infinity Principle. In the subatomic realm, for example, physicists can no longer think of an electron as a classical particle following a smooth path in the same way that a planet or a cannonball does. According to quantum mechanics, trajectories become jittery, blurry, and poorly defined at the microscopic scale, so we need to describe the behavior of electrons as probability waves instead of Newtonian trajectories. As soon as we do that, however, calculus returns triumphantly. It governs the evolution of probability waves through something called the Schrödinger equation.

It's incredible but true: Even in the subatomic realm where Newtonian physics breaks down, Newtonian calculus still works. In fact, it works spectacularly well. As we'll see in the pages ahead, it has teamed up with quantum mechanics to predict the remarkable effects that underlie medical imaging, from MRI and CT scans to the more exotic positron emission tomography.

It's time for us to take a closer look at the language of the universe. Naturally, the place to start is at infinity.

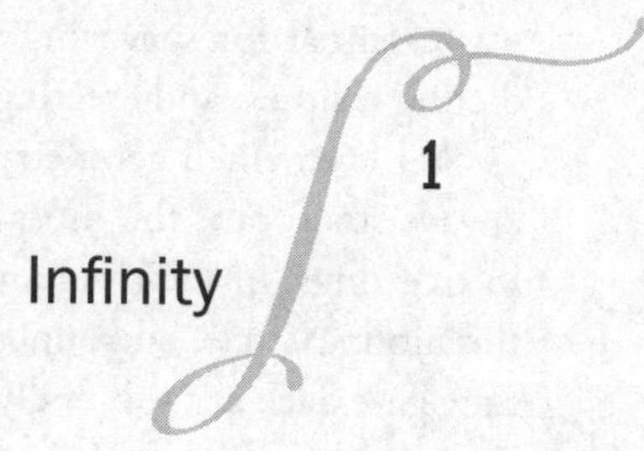

1 Infinity

The beginnings of mathematics were grounded in everyday concerns. Shepherds needed to keep track of their flocks. Farmers needed to weigh the grain reaped in the harvest. Tax collectors had to decide how many cows or chickens each peasant owed the king. Out of such practical demands came the invention of numbers. At first they were tallied on fingers and toes. Later they were scratched on animal bones. As their representation evolved from scratches to symbols, numbers facilitated everything from taxation and trade to accounting and census taking. We see evidence of all this in Mesopotamian clay tablets written more than five thousand years ago: row after row of entries recorded with the wedge-shaped symbols called cuneiform.

Along with numbers, shapes mattered too. In ancient Egypt, the measurement of lines and angles was of paramount importance. Each year surveyors had to redraw the boundaries of farmers' fields after the summer flooding of the Nile washed the borderlines away. That activity later gave its name to the study of shape in general: *geometry*, from the Greek *gē*, "earth," and *metrēs*, "measurer."

At the start, geometry was hard-edged and sharp-cornered. Its predilection for straight lines, planes, and angles reflected its utilitarian origins—triangles were useful as ramps, pyramids as monuments and tombs, and rectangles as tabletops, altars, and plots of land. Builders and carpenters used right angles for plumb lines. For sailors, architects, and priests, knowledge of straight-line geometry was essential for surveying, navigating, keeping the calendar, predicting eclipses, and erecting temples and shrines.

Yet even when geometry was fixated on straightness, one curve always stood out, the most perfect of all: the circle. We see circles in tree rings, in the ripples on a pond, in the shape of the sun and the moon. Circles surround us in nature. And as we gaze at circles, they gaze back at us, literally. There they are in the eyes of our loved ones, in the circular outlines of their pupils and irises. Circles span the practical and the emotional, as wheels and wedding rings, and they are mystical too. Their eternal return suggests the cycle of the seasons, reincarnation, eternal life, and never-ending love. No wonder circles have commanded attention for as long as humanity has studied shapes.

Mathematically, circles embody change without change. A point moving around the circumference of a circle changes direction without ever changing its distance from a center. It's a minimal form of change, a way to change and curve in the slightest way possible. And, of course, circles are symmetrical. If you rotate a circle about its center, it looks unchanged. That rotational symmetry may be why circles are so ubiquitous. Whenever some aspect of nature doesn't care about direction, circles are bound to appear. Consider what happens when a raindrop hits a puddle: tiny ripples expand outward from the point of impact. Because they spread equally fast in all directions and because they started at a single point, the ripples *have* to be circles. Symmetry demands it.

Circles can also give birth to other curved shapes. If we imagine skewering a circle on its diameter and spinning it around that axis in three-dimensional space, the rotating circle makes a sphere, the shape of a globe or a ball. When a circle is moved vertically into the third dimension along a straight line at right angles to its plane, it makes a cylinder, the shape of a can or a hatbox. If it shrinks at the same time as it's moving vertically, it makes a cone; if it expands as it moves vertically, it makes a truncated cone (the shape of a lampshade).

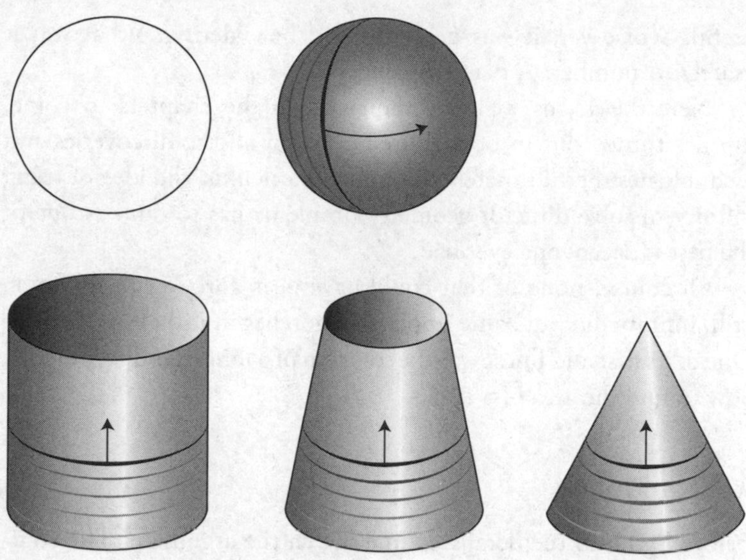

Circles, spheres, cylinders, and cones fascinated the early geometers, but they found them much harder to analyze than triangles, rectangles, squares, cubes, and other rectilinear shapes made of straight lines and flat planes. They wondered about the areas of curved surfaces and the volumes of curved solids but had no clue how to solve such problems. Roundness defeated them.

Infinity as a Bridge Builder

Calculus began as an outgrowth of geometry. Back around 250 BCE in ancient Greece, it was a hot little mathematical startup devoted to the mystery of curves. The ambitious plan of its devotees was to use infinity to build a bridge between the curved and the straight. The hope was that once that link was established, the methods and techniques of straight-line geometry could be shuttled across the bridge and brought to bear on the mystery of curves. With infinity's help, all the old problems could be solved. At least, that was the pitch.

At the time, that plan must have seemed pretty far-fetched. Infinity had a dubious reputation. It was known for being scary, not useful. Worse yet, it was nebulous and bewildering. What was it exactly? A number? A place? A concept?

Nevertheless, as we'll see soon and in the chapters to come, infinity turned out to be a godsend. Given all the discoveries and technologies that ultimately flowed from calculus, the idea of using infinity to solve difficult geometry problems has to rank as one of the best ideas anyone ever had.

Of course, none of that could have been foreseen in 250 BCE. Still, infinity did put some impressive notches in its belt right away. One of its first and finest was the solution of a long-standing enigma: how to find the area of a circle.

A Pizza Proof

Before I go into the details, let me sketch the argument. The strategy is to reimagine the circle as a pizza. Then we'll slice that pizza into infinitely many pieces and magically rearrange them to make a rectangle. That will give us the answer we're looking for, since moving slices around obviously doesn't change their area from what they were originally, and we know how to find the area of a rectangle: we just multiply its width times its height. The result is a formula for the area of a circle.

For the sake of this argument, the pizza needs to be an idealized mathematical pizza, perfectly flat and round, with an infinitesimally thin crust. Its circumference, abbreviated by the letter *C*, is the distance around the pizza, measured by tracing around the crust. Circumference isn't something that pizza lovers ordinarily care about, but if we wanted to, we could measure *C* with a tape measure.

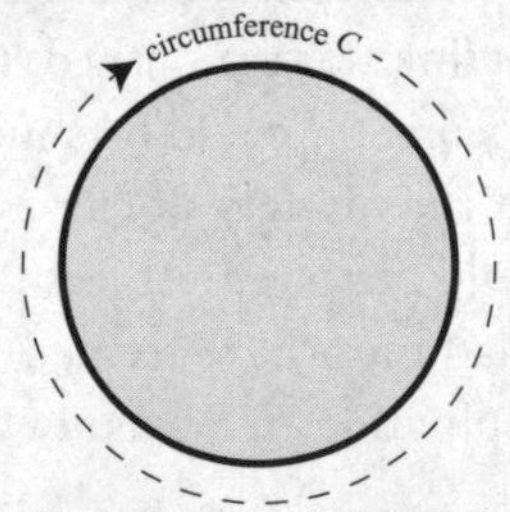

Another quantity of interest is the pizza's radius, r, defined as the distance from its center to every point on its crust. In particular, r also measures how long the straight side of a slice is, assuming that all the slices are equal and cut from the center out to the crust.

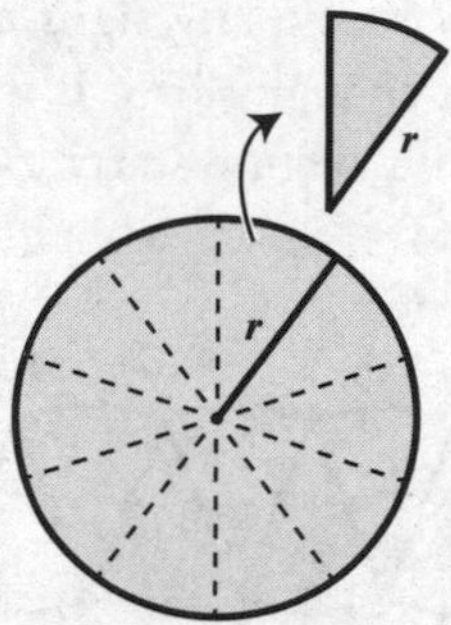

Suppose we start by dividing the pie into four quarters. Here's one way to rearrange them, but it doesn't look too promising.

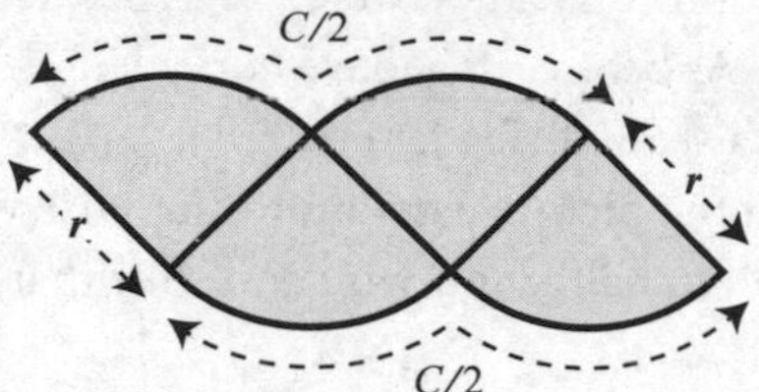

The new shape looks bulbous and strange with its scalloped top and bottom. It's certainly not a rectangle, so its area is not easy to guess. We seem to be going backward. But as in any drama, the hero needs to get into trouble before triumphing. The dramatic tension is building.

While we're stuck here, though, we should notice two things, because they are going to hold true throughout the proof, and they will ultimately give us the dimensions of the rectangle we're seeking. The first observation is that half of the crust became the curvy top of the new shape, and the other half became the bottom. So the curvy top has a length equal to half the circumference, $C/2$, and so does the bottom, as shown in the diagram. That length is eventually going to turn into the long side of the rectangle, as we'll see. The other thing to notice is that the tilted straight sides of the bulbous shape are just the sides of the original pizza slices, so they still have length r. That length is eventually going to turn into the short side of the rectangle.

The reason we aren't seeing any signs of the desired rectangle yet is that we haven't cut enough slices. If we make eight slices and rearrange them like so, our picture starts to look more nearly rectangular.

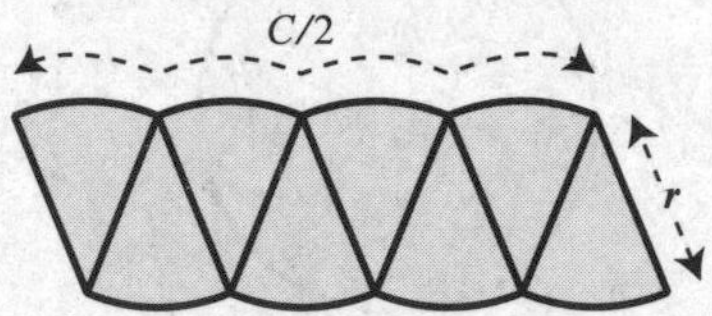

In fact, the pizza starts to look like a parallelogram. Not bad—at least it's almost rectilinear. And the scallops on the top and bottom are a lot less bulbous than they were. They flattened out when we used more slices. As before, they have curvy length $C/2$ on the top and bottom and a slanted-side length r.

To spruce up the picture even more, suppose we cut one of the slanted end pieces in half lengthwise and shift that half to the other side.

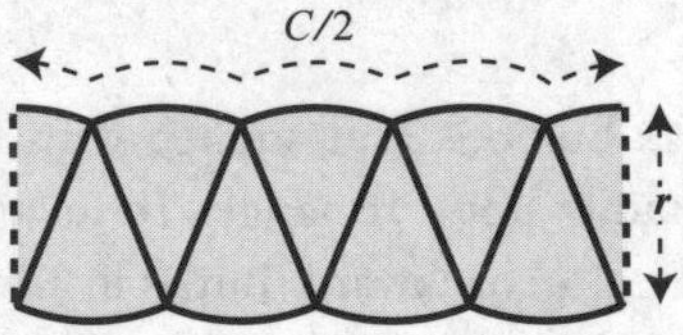

Now the shape looks very much like a rectangle. Admittedly, it's still not perfect because of the scalloped top and bottom caused by the curvature of the crust, but at least we're making progress.

Since making more pieces seems to be helping, let's keep slicing. With sixteen slices and the cosmetic sprucing-up of the end piece, as we did before, we get this result:

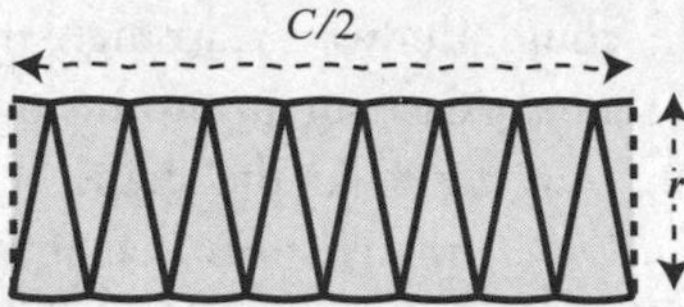

The more slices we take, the more we flatten out the scallops produced by the crust. Our maneuvers are producing a sequence of shapes that are magically homing in on a certain rectangle. Because the shapes keep getting closer and closer to that rectangle, we'll call it the *limiting* rectangle.

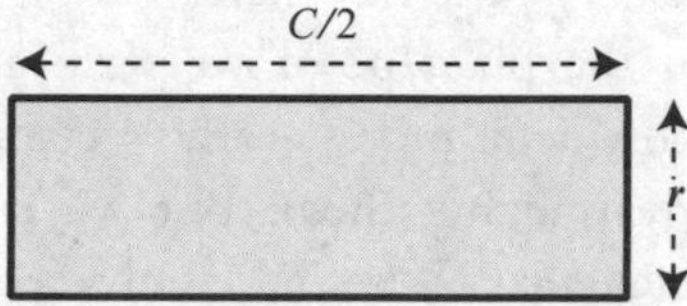

The point of all this is that we can easily find the area of this limiting rectangle by multiplying its width by its height. All that remains is to find that height and width in terms of the circle's dimensions. Well, since the slices are standing upright, the height is just the radius r of the original circle. And the width is half the circumference of the circle; that's because half of the circumference (the crust of the pizza) went into making the top of the rectangle and the other half got used on the bottom, just as it did at every intermediate stage of working with the bulbous shapes. Thus the width is half the circumference, $C/2$. Putting everything together, the area of the limiting rectangle is given by its height times its width, namely, $A = r \times C/2 = rC/2$. And since moving the pizza slices around did not change their area, this must also be the area of the original circle!

This result for the area of a circle, $A = rC/2$, was first proved (using a similar but much more careful argument) by the ancient Greek mathematician Archimedes (287–212 BCE) in his essay "Measurement of a Circle."

The most innovative aspect of the proof is the way infinity came to the rescue. When we had only four slices, or eight, or sixteen, the best we could do was rearrange the pizza into an imperfect scalloped shape. After an unpromising start, the more slices we took, the more rectangular the shape became. But it was only in the limit of *infinitely* many slices that it became truly rectangular. That's the big idea behind calculus. Everything becomes simpler at infinity.

Limits and the Riddle of the Wall

A limit is like an unattainable goal. You can get closer and closer to it, but you can never get all the way there.

For example, in the pizza proof we were able to make the scalloped shapes more and more nearly rectangular by cutting enough slices and rearranging them. But we could never make them genuinely rectangular. We could only approach that state of perfection. Fortunately, in calculus, the unattainability of the limit usually doesn't matter. We can often solve the problems we're working on by fantasizing that we can actually reach the limit and then seeing what that fantasy implies. In fact, many of the greatest pioneers of the subject did precisely that and made great discoveries by doing so. Logical, no. Imaginative, yes. Successful, very.

A limit is a subtle concept but a central one in calculus. It's elusive because it's not a common idea in daily life. Perhaps the closest analogy is the Riddle of the Wall. If you walk halfway to the wall, and then you walk half the remaining distance, and then you walk half of that, and on and on, will there ever be a step when you finally get to the wall?

The answer is clearly no, because the Riddle of the Wall stipulates that at each step, you walk halfway to the wall, not all the way. After you take ten steps or a million or any other number of steps, there will always be a gap between you and the wall. But equally clearly, you can get arbitrarily close to the wall. What this means is that by taking enough steps, you can get to within a centimeter of it, or a millimeter, or a nanometer, or any other tiny but nonzero distance, but you can never get all the way there. Here, the wall plays the role of the limit. It took about two thousand years for the limit concept to be rigorously defined. Until then, the pioneers of calculus got by just fine with intuition. So don't worry if limits feel hazy for now. We'll get to know them better by watching them in action. From a modern perspective, they matter because they are the bedrock on which all of calculus is built.

If the metaphor of the wall seems too bleak and inhuman (who wants to approach a wall?), try this analogy: Anything that approaches a limit is like a hero engaged in an endless quest. It's not an exercise in total futility, like the hopeless task faced by Sisyphus, who was condemned to roll a boulder up a hill only to see it roll back down again over and over for eternity. Rather, when a mathematical process advances toward a limit (like the scalloped shapes homing in on the limiting rectangle), it's as if a protagonist is striving for something he knows is impossible but for which he still holds out the hope of success, encouraged by the steady progress he's making while trying to reach an unreachable star.

The Parable of .333 . . .

To reinforce the big ideas that everything becomes simpler at infinity and that limits are like unattainable goals, consider the following example from arithmetic. It's the problem of converting a fraction—for example, ⅓—into an equivalent decimal (in this case, ⅓ = 0.333 . . .). I vividly remember when my eighth-grade math teacher, Ms. Stanton, taught us how to do this. It was memorable because she suddenly started talking about infinity.

Until that moment, I'd never heard a grownup mention infinity. My parents certainly had no use for it. It seemed like a secret that only kids knew about. On the playground, it came up all the time in taunts and one-upmanship.

"You're a jerk!"
"Yeah, well, you're a jerk times two!"
"And you're a jerk times infinity!"
"And you're a jerk times infinity plus one!"
"That's the same as infinity, you idiot!"

Those edifying sessions had convinced me that infinity did not behave like an ordinary number. It didn't get bigger when you added one to it. Even adding infinity to it didn't help. Its invincible properties made it great for finishing arguments in the schoolyard. Whoever deployed it first would win.

But no teacher had ever talked about infinity until Ms. Stanton brought it up that day. Everyone in our class already knew about finite decimals, the familiar kind used for amounts of money, like $10.28, with its two digits after the decimal point. By comparison, infinite decimals, which had infinitely many digits after the decimal point, seemed strange at first but appeared natural as soon as we started to discuss fractions.

We learned that the fraction ⅓ could be written as 0.333 . . . where the dot-dot-dots meant that the threes repeated indefinitely. That made sense to me, because when I tried to calculate ⅓ by doing the long-division algorithm on it, I found myself stuck in an endless loop: three doesn't go into one, so pretend the one is a ten; then three goes into ten three times, which leaves a remainder of one; and now I'm back where I started, still trying to divide three into one. There was no way out of the loop. That's why the threes kept repeating in 0.333 . . . .

The three dots at the end of 0.333 . . . have two interpretations. The naive interpretation is that there are literally infinitely many 3s packed side by side to the right of the decimal point. We can't write them all down, of course, since there are infinitely many of them, but by writing the three dots we signify that they are all there, at least in our minds. I'll call this the *completed infinity* interpretation. The advantage of this interpretation is that it seems easy and commonsensical, as long as we are willing not to think too hard about what infinity means.

The more sophisticated interpretation is that 0.333 . . . represents a limit, just like the limiting rectangle does for the scalloped shapes in the pizza proof or like the wall does for the hapless walker. Except here, 0.333 . . . represents the limit of the successive decimals we generate by doing long division on the fraction ⅓. As the division process continues for more and more steps, it generates more and more 3s in the decimal expansion of ⅓. By grinding away, we can produce an approximation as close to ⅓ as we like. If we're not happy with $1/3 \approx 0.3$, we can always go a step further to $1/3 \approx 0.33$, and so on. I'll call this the *potential infinity* interpretation. It's "potential" in the sense that the approximations can potentially go on for as long as desired. There's nothing to stop us from continuing for a million or a billion or any other number of steps. The advantage of this interpretation is that we never have to invoke woolly-headed notions like infinity. We can stick to the finite.

For working with equations like $1/3 = 0.333\ldots$, it doesn't really matter which view we take. They're equally tenable and yield the same mathematical results in any calculation we care to perform. But there are other situations in mathematics where the completed infinity interpretation can cause logical mayhem. This is what I meant in the introduction when I raised the specter of the golem of infinity. Sometimes it really does make a difference how we think about the results of a process that approaches a limit. Pretending that the process actually terminates and that it somehow reaches the nirvana of infinity can occasionally get us into trouble.

The Parable of the Infinite Polygon

As a chastening example, suppose we put a certain number of dots on a circle, space them evenly, and connect them to one another with straight lines. With three dots, we get an equilateral triangle; with four, a square; with five, a pentagon; and so on, running through a sequence of rectilinear shapes called regular polygons.

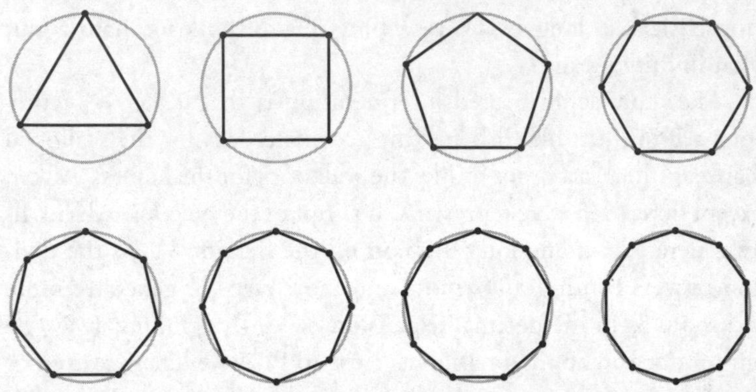

Notice that the more dots we use, the rounder the polygons become and the closer they get to the circle. Meanwhile, their sides get shorter and more numerous. As we move progressively further through the sequence, the polygons approach the original circle as a limit.

In this way, infinity is bridging two worlds again. This time it's taking us from the rectilinear to the round, from sharp-cornered polygons to silky-smooth circles, whereas in the pizza proof, infinity brought us from round to rectilinear as it transformed a circle into a rectangle.

Of course, at any finite stage, a polygon is still just a polygon. It's not yet a circle and it never becomes one. It gets closer and closer to being a circle, but it never truly gets there. We are dealing here with potential infinity, not completed infinity. So everything is airtight from the standpoint of logical rigor.

But what if we could go all the way to completed infinity? Would the resulting infinite polygon with infinitesimally short sides actually *be* a circle? It's tempting to think so, because then the polygon would be smooth. All its corners would be sanded off. Everything would become perfect and beautiful.

The Allure and Peril of Infinity

There's a general lesson here: Limits are often simpler than the approximations leading up to them. A circle is simpler and more graceful than any of the thorny polygons that approach it. So too for the pizza proof, where the limiting rectangle was simpler and more elegant than the scalloped shapes, with their unsightly bulges and cusps. And likewise for the fraction $1/3$. It was simpler and more handsome than any of the ungainly fractions creeping up on it, with their big ugly numerators and denominators, like $3/10$ and $33/100$ and $333/1000$. In all these cases, the limiting shape or number was simpler and more symmetrical than its finite approximators.

This is the allure of infinity. Everything becomes better there.

With that lesson in mind, let's return to the parable of the infinite polygon. Should we take the plunge and say that a circle truly *is* a polygon with infinitely many infinitesimal sides? No. We mustn't do that, mustn't yield to that temptation. Doing so would be to commit the sin of completed infinity. It would condemn us to logical hell.

To see why, suppose we entertain the thought, just for a moment, that a circle is indeed an infinite polygon with infinitesimal sides. How long, exactly, are those sides? Zero length? If so, then infinity times zero—the combined length of all those sides—must equal the circumference of the circle. But now imagine a circle of double the circumference. Infinity times zero would also have to equal that larger circumference as well. So infinity times zero would have to be both the circumference and double the circumference. What nonsense! There simply is no consistent way to define infinity times zero, and so there is no sensible way to regard a circle as an infinite polygon.

Nevertheless, there is something so enticing about this intuition. Like the biblical original sin, the original sin of calculus—the temptation to treat a circle as an infinite polygon with infinitesimally short sides—is very hard to resist, and for the same reason. It tempts us with the prospect of forbidden knowledge, with insights unavailable by ordinary means. For thousands of years, geometers struggled to figure out the circumference of a circle. If only a circle could be replaced by a polygon made of many tiny straight sides, the problem would be so much easier.

By listening to the hiss of this serpent—but holding back just enough, by using potential infinity instead of the more tempting completed infinity—mathematicians learned how to solve the circumference problem and other mysteries of curves. In the coming chapters, we'll see how they did it. But first, we need to gain an even deeper appreciation of just how dangerous completed infinity can be. It's a gateway sin to many others, including the sin our teachers warned us about first.

The Sin of Dividing by Zero

All across the world, students are being taught that division by zero is forbidden. They should feel shocked that such a taboo exists. Numbers are supposed to be orderly and well behaved. Math class is a place for logic and reasoning. And yet it's possible to ask simple things of numbers that just don't work or make sense. Dividing by zero is one of them.

The root of the problem is infinity. Dividing by zero summons infinity in much the same way that a Ouija board supposedly summons spirits from another realm. It's risky. Don't go there.

For those who can't resist and want to understand why infinity lurks in the shadows, imagine dividing 6 by a number that's small and getting close to zero, but that isn't quite zero, say something like 0.1. There's nothing taboo about that. The answer to 6 divided by 0.1 is 60, a fairly sizable number. Divide 6 by an even smaller number, say 0.01, and the answer grows bigger; now it's 600. If we dare to divide 6 by a number much closer to zero, say 0.0000001, the answer gets much bigger; instead of 60 or 600, now it's 60,000,000. The trend is clear. The smaller the divisor, the bigger the answer. In the limit as the divisor approaches zero, the answer approaches infinity. That's the real reason why we can't divide by zero. The faint of heart say the answer is undefined, but the truth is it's infinite.

All of this can be visualized as follows. Imagine dividing a 6-centimeter line into pieces that are each 0.1 centimeter long. Those 60 pieces laid end to end make up the original.

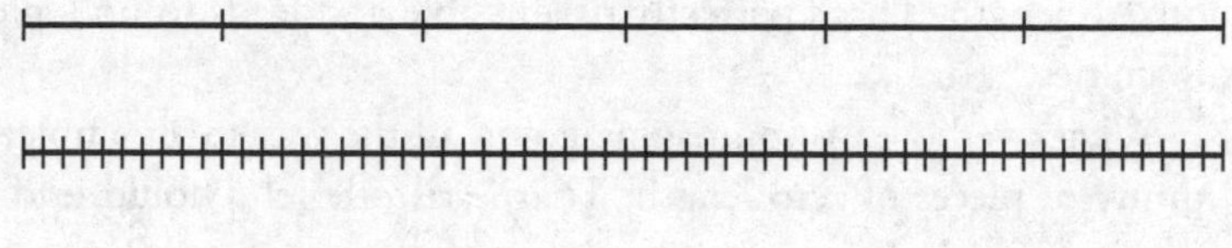

0.1

Likewise (but I won't attempt to sketch it), that same line can be chopped into 600 pieces that are each 0.01 centimeter or 60,000,000 pieces that are each 0.0000001 centimeter.

If we keep going and take this chopping frenzy to the limit, we are led to the bizarre conclusion that a 6-centimeter line is made up of *infinitely* many pieces of length *zero*. Maybe that sounds plausible. After all, the line is made up of infinitely many points, and each point has zero length.

But what's so philosophically unnerving is that the same argument applies to a line of *any* length. Indeed, there's nothing special about the number 6. We could just as well have claimed that a line of length 3 centimeters, or 49.57, or 2,000,000,000 is made up of infinitely many points of zero length. Evidently, multiplying zero by infinity can give us any and every conceivable result—6 or 3 or 49.57 or 2,000,000,000. That's horrifying, mathematically speaking.

The Sin of Completed Infinity

The transgression that dragged us into this mess was pretending that we could actually *reach* the limit, that we could treat infinity like an attainable number. Back in the fourth century BCE, the Greek philosopher Aristotle warned that sinning with infinity in this way could lead to all sorts of logical trouble. He railed against what he called completed infinity and argued that only potential infinity made sense.

In the context of chopping a line into pieces, potential infinity would mean that the line could be cut into more and more pieces, as many as desired but still always a finite number and all of nonzero length. That's perfectly permissible and leads to no logical difficulties.

What's verboten is to imagine going all the way to a completed infinity of pieces of zero length. That, Aristotle felt, would lead to nonsense—as it does here, in revealing that zero times infinity can give any answer. And so he forbade the use of completed infinity in mathematics and philosophy. His edict was upheld by mathematicians for the next twenty-two hundred years.

Somewhere in the dark recesses of prehistory, somebody realized that numbers never end. And with that thought, infinity was born. It's the numerical counterpart of something deep in our psyches, in our nightmares of bottomless pits, and in our hopes for eternal life. Infinity lies at the heart of so many of our dreams and fears and unanswerable questions: How big is the universe? How long is forever? How powerful is God? In every branch of human thought, from religion and philosophy to science and mathematics, infinity has befuddled the world's finest minds for thousands of years. It has been banished, outlawed, and shunned. It's always been a dangerous idea. During the Inquisition, the renegade monk Giordano Bruno was burned alive at the stake for suggesting that God, in His infinite power, created innumerable worlds.

Zeno's Paradoxes

About two millennia before the execution of Giordano Bruno, another brave philosopher dared to contemplate infinity. Zeno of Elea (c. 490–430 BCE) posed a series of paradoxes about space, time, and motion in which infinity played a starring and perplexing role. These conundrums anticipated ideas at the heart of calculus and are still being debated today. Bertrand Russell called them immeasurably subtle and profound.

We aren't sure what Zeno was trying to prove with his paradoxes because none of his writings have survived, if any existed to begin with. His arguments have come down to us through Plato and Aristotle, who summarized them mainly to demolish them. In their telling, Zeno was trying to prove that change is impossible. Our senses tell us otherwise, but our senses deceive us. Change, according to Zeno, is an illusion.

Three of Zeno's paradoxes are particularly famous and strong. The first of them, the Paradox of the Dichotomy, is similar to the Riddle of the Wall but vastly more frustrating. It holds that you can't ever move because before you can take a single step, you need to take a half a step. And before you can do that, you need to take a quarter of a step, and so on. So not only can't you get to the wall—you can't even start walking.

It's a brilliant paradox. Who would have thought that taking a step required completing infinitely many subtasks? Worse still, there is no *first* task to complete. The first task cannot be taking half a step because before that you'd have to complete a quarter of a step, and before that, an eighth of a step, and so on. If you thought you had a lot to do before breakfast, imagine having to finish an infinite number of tasks just to get to the kitchen.

Another paradox, called Achilles and the Tortoise, maintains that a swift runner (Achilles) can never catch up to a slow runner (a tortoise) if the slow runner has been given a head start in a race.

For by the time Achilles reaches the spot where the tortoise started, the tortoise will have moved a little bit farther down the track. And by the time Achilles reaches that new location, the tortoise will have crept slightly farther ahead. Since we all believe that a fast runner *can*

overtake a slow runner, either our senses are deceiving us or there is something wrong in the way that we reason about motion, space, and time.

In these first two paradoxes, Zeno seemed to be arguing against space and time being fundamentally continuous, meaning that they can be divided endlessly. His clever rhetorical strategy (some say he invented it) was proof by contradiction, known to lawyers and logicians as *reductio ad absurdum*, reduction to an absurdity. In both paradoxes, Zeno assumed the continuity of space and time and then deduced a contradiction from that assumption; therefore, the assumption of continuity must be false. Calculus is founded on that very assumption and so has a lot at stake in this fight. It rebuts Zeno by showing where his reasoning went wrong.

For example, here's how calculus takes care of Achilles and the tortoise. Suppose the tortoise starts 10 meters ahead of Achilles but Achilles runs 10 times faster, say at a speed of 10 meters per second compared to the tortoise's 1 meter per second. Then it takes Achilles 1 second to make up the tortoise's 10-meter head start. During that time the tortoise will have moved 1 meter farther ahead. It takes Achilles another 0.1 second to make up that difference, by which time the tortoise will have moved another 0.1 meter ahead. Continuing this reasoning, we see that Achilles's consecutive catch-up times add up to the infinite series

$$1 + 0.1 + 0.01 + 0.001 + \cdots = 1.111\ldots \text{ seconds.}$$

Rewritten as an equivalent fraction, this amount of time is equal to $10/9$ seconds. That's how long it takes Achilles to catch up to the tortoise and overtake him. And although Zeno was right that Achilles has infinitely many tasks to complete, there's nothing paradoxical about that. As the math shows, he can do them all in a finite amount of time.

This line of reasoning qualifies as a calculus argument. We just summed an infinite series and calculated a limit, as we did earlier when we discussed why $0.333\ldots = 1/3$. Whenever we work with infinite decimals, we are doing calculus (even though most people would pooh-pooh it as middle-school arithmetic).

Incidentally, calculus isn't the only way to solve this problem. We could use algebra instead. To do so, we first need to figure out where each runner is on the track at an arbitrary time t seconds after the race begins. Since Achilles runs at a speed of 10 meters per second and since distance equals rate times time, his distance down the track is $10t$. As for the tortoise, he had a head start of 10 meters and he runs with a speed of 1 meter per second, so his distance down the track is $10 + t$. To ascertain the time when Achilles overtakes the tortoise, we have to set those two expressions equal to one another, because that's the algebraic way of asking when Achilles and the tortoise are at the same place at the same time. The resulting equation is

$$10t = 10 + t.$$

To solve this equation, subtract t from both sides. That gives $9t = 10$. Then divide both sides by 9. The result, $t = 10/9$ seconds, is the same as we found with infinite decimals.

So from the perspective of calculus, there really is no paradox about Achilles and the tortoise. If space and time are continuous, everything works out nicely.

Zeno Goes Digital

In a third paradox, the Paradox of the Arrow, Zeno argued against an alternative possibility—that space and time are fundamentally discrete, meaning that they are composed of tiny indivisible units, something like pixels of space and time. The paradox goes like this. If space and time are discrete, an arrow in flight can never move, because at each instant (a pixel of time) the arrow is at some definite place (a specific set of pixels in space). Hence, at any given instant, the arrow is not moving. It is also not moving between instants because, by assumption, there *is* no time between instants. Therefore, at no time is the arrow ever moving.

To my mind, this is the most subtle and interesting of the paradoxes. Philosophers are still debating its status, but it seems to me that Zeno got it two-thirds right. In a world where space and time are discrete, an arrow in flight *would* behave as Zeno said. It would strangely materialize at one place after another as time clicks forward in discrete steps. And he was also right that our senses tell us that the real world is not like that, at least not as we ordinarily perceive it.

But Zeno was wrong that motion would be impossible in such a world. We all know this from our experience of watching movies and videos on our digital devices. Our cell phones and DVRs and computer screens chop everything into discrete pixels, and yet, contrary to Zeno's assertion, motion can take place perfectly well in these discretized landscapes. As long as everything is diced fine enough, we can't tell the difference between a smooth motion and its digital representation. If we were to watch a high-resolution video of an arrow in flight, we'd actually be seeing a pixelated arrow materializing in one discrete frame after another. But because of our perceptual limitations, it would look like a smooth trajectory. Sometimes our senses really do deceive us.

Of course, if the chopping is too blocky, we *can* tell the difference between the continuous and the discrete, and we often find it bothersome. Consider how an old-fashioned analog clock differs from a modern-day digital/mechanical monstrosity. On the analog clock, the second hand sweeps around in a beautifully uniform motion. It depicts time as flowing. Whereas on the digital clock, the second hand jerks forward in discrete steps, *thwack, thwack, thwack*. It depicts time as jumping.

Infinity can build a bridge between these two very different conceptions of time. Imagine a digital clock that advances through trillions of little clicks per second instead of one loud *thwack*. We would no longer be able to tell the difference between that kind of digital clock and a true analog clock. Likewise with movies and videos; as long as the frames flash by fast enough, say at thirty frames a second, they give the impression of seamless flow. And if there were *infinitely* many frames per second, the flow truly would be seamless.

Consider how music is recorded and played back. My younger

daughter recently received an old-fashioned Victrola record player for her fifteenth birthday. She's now able to listen to Ella Fitzgerald on vinyl. This is a quintessential analog experience. All of Ella's notes and scats glide just as smoothly as they did when she sang them; her volume goes continuously from soft to loud and everywhere in between, and her pitch climbs just as gracefully from low to high. Whereas when you listen to her on digital, every aspect of her music is minced into tiny, discrete steps and converted into strings of 0s and 1s. Although conceptually the differences are gigantic, our ears can't hear them.

So in everyday life, the gulf between the discrete and the continuous can often be bridged, at least to a good approximation. For many practical purposes, the discrete can stand in for the continuous, as long as we slice things thinly enough. In the ideal world of calculus, we can go one better. Anything that's continuous can be sliced *exactly* (not just approximately) into infinitely many infinitesimal pieces. That's the Infinity Principle. With limits and infinity, the discrete and the continuous become one.

Zeno Meets the Quantum

The Infinity Principle asks us to pretend that everything can be sliced and diced endlessly. We've already seen how useful such concepts can be. Imagining pizzas that can be cut into arbitrarily thin pieces enabled us to find the area of a circle exactly. The question naturally arises: Do such infinitesimally small things exist in the real world?

Quantum mechanics has something to say about that. It's the branch of modern physics that describes how nature behaves at its smallest scales. It's the most accurate physical theory ever devised, and it is legendary for its weirdness. Its terminology, with its zoo of leptons, quarks, and neutrinos, sounds like something out of Lewis Carroll. The behavior it describes is often weird as well. At the atomic scale, things can happen that would never occur in the macroscopic world.

For instance, consider the Riddle of the Wall from a quantum perspective. If the walker were an electron, there's a chance it might walk right through the wall. This effect is known as quantum tunneling. It actually occurs. It's hard to make sense of this in classical terms, but the quantum explanation is that electrons are described by probability waves. Those waves obey an equation formulated in 1925 by the Austrian physicist Erwin Schrödinger. The solution to Schrödinger's equation shows that a small portion of the electron probability wave exists on the far side of an impenetrable barrier. This means there is some small but nonzero probability that the electron will be detected on the far side of the barrier, as if it had tunneled through the wall. With the help of calculus, we can calculate the rate at which such tunneling events occur, and experiments have confirmed the predictions. Tunneling is real. Alpha particles tunnel out of uranium nuclei at the predicted rate to produce the effect known as radioactivity. Tunneling also plays an important role in the nuclear-fusion processes that make the sun shine, so life on Earth depends partially on tunneling. And it has many technological uses; scanning tunneling microscopy, which allows scientists to image and manipulate individual atoms, is based on the concept.

We have no intuition for such events at the atomic scale, being the gargantuan creatures composed of trillions upon trillions of atoms that we are. Fortunately, calculus can take the place of intuition. By applying calculus and quantum mechanics, physicists have opened a theoretical window on the microworld. The fruits of their insights include lasers and transistors, the chips in our computers, and the LEDs in our flat-screen TVs.

Although quantum mechanics is conceptually radical in many respects, in Schrödinger's formulation, it retains the traditional assumption that space and time are continuous. Maxwell made the same assumption in his theory of electricity and magnetism; so did Newton in his theory of gravity and Einstein in his theory of relativity. All of calculus, and hence all of theoretical physics, hinges on this assumption of continuous space and time. That assumption of continuity has been resoundingly successful so far.

But there is reason to believe that at much, much smaller scales of the universe, far below the atomic scale, space and time may ultimately lose their continuous character. We don't know for sure what it's like down there, but we can guess. Space and time might become as neatly pixelated as Zeno imagined in his Paradox of the Arrow, but more likely they'd degenerate into a disorderly mess because of quantum uncertainty. At such small scales, space and time might seethe and roil at random. They might fluctuate like bubbling foam.

Although there is no consensus about how to visualize space and time at these ultimate scales, there is universal agreement about how small those scales are likely to be. They are forced upon us by three fundamental constants of nature. One of them is the gravitational constant, G. It measures the strength of gravity in the universe. It appeared first in Newton's theory of gravity and again in Einstein's general theory of relativity. It is bound to occur in any future theory that supersedes them. The second constant, $\hbar$ (pronounced "h bar"), reflects the strength of quantum effects. It appears, for example, in Heisenberg's uncertainty principle and in Schrödinger's wave equation of quantum mechanics. The third constant is the speed of light, c. It is the speed limit for the universe. No signal of any kind can travel faster than c. This speed must necessarily enter any theory of space and time because it ties the two of them together via the principle that distance equals rate times time, where c plays the role of the rate or speed.

In 1899, the father of quantum theory, a German physicist named Max Planck, realized that there was one and only one way to combine these fundamental constants to produce a scale of length. That unique length, he concluded, was a natural yardstick for the universe. In his honor, it is now called the Planck length. It is given by the algebraic combination

$$\text{Planck length} = \sqrt{\frac{\hbar G}{c^3}}\,.$$

When we plug in the measured values of G, $\hbar$, and c, the Planck length comes out to be about 10^{-35} meters, a stupendously small distance that's about a hundred million trillion times smaller than the diameter of a proton. The corresponding Planck time is the time it would take light to traverse this distance, which is about 10^{-43}

seconds. Space and time would no longer make sense below these scales. They're the end of the line.

These numbers put a bound on how fine we could ever slice space or time. To get a feel for the level of precision we're talking about here, consider how many digits we would need to make one of the most extreme comparisons imaginable. Take the largest possible distance, the estimated diameter of the known universe, and divide it by the smallest possible distance, the Planck length. That unfathomably extreme ratio of distances is a number with only sixty digits in it. I want to stress that—*only* sixty digits. That's the most we would ever need to express one distance in terms of another. Using more digits than that—say a hundred digits, let alone infinitely many—would be colossal overkill, *way* more than we would ever need to describe any real distances out there in the material world.

And yet in calculus, we use infinitely many digits all the time. As early as middle school, students are asked to think about numbers like 0.333 . . . whose decimal expansion goes on forever. We call these real numbers, but there is nothing real about them. The requirement to specify a real number by an infinite number of digits after the decimal point is exactly what it means to be *not* real, at least as far as we understand reality through physics today.

If real numbers are not real, why do mathematicians love them so much? And why are schoolchildren forced to learn about them? Because calculus needs them. From the beginning, calculus has stubbornly insisted that everything—space and time, matter and energy, all objects that ever have been or will be—should be regarded as continuous. Accordingly, everything can and should be quantified by real numbers. In this idealized, imaginary world, we pretend that everything can be split finer and finer without end. The whole theory of calculus is built on that assumption. Without it, we couldn't compute limits, and without limits, calculus would come to a clanking halt. If all we ever used were decimals with only sixty digits of precision, the number line would be pockmarked and cratered. There would be holes where pi, the square root of two, and any other numbers that need infinitely many digits after the decimal point should exist. Even a simple fraction such as ⅓ would be missing, because it too requires an infinite number of digits (0.333 . . .) to pinpoint its location on the number line. If we want to think of the totality of all numbers as forming a continuous line, those numbers have to be real numbers. They may be an approximation of reality, but they work amazingly well. Reality is too hard to model any other way. With infinite decimals, as with the rest of calculus, infinity makes everything simpler.

2

The Man Who Harnessed Infinity

ABOUT TWO HUNDRED years after Zeno pondered the nature of space, time, motion, and infinity, another thinker found infinity irresistible. His name was Archimedes. We've met him already in connection with the area of a circle, but he is legendary for many other reasons.

For one thing, there are a lot of funny stories about him. Several portray him as the original math geek. For example, the historian Plutarch tells us that Archimedes could become so engrossed in geometry that it "made him forget his food and neglect his person." (That certainly rings true. For many of us mathematicians, meals and personal hygiene aren't top priorities.) Plutarch goes on to say that when Archimedes was lost in his mathematics, he would have to be "carried by absolute violence to bathe." It's interesting that he was such a reluctant bather, given that a bath is the setting for the one story about him that everybody knows. According to the Roman architect Vitruvius, Archimedes became so excited by a sudden insight he had in the bath that he leaped out of the tub and ran down the street naked shouting, "Eureka!" ("I have found it!")

Other stories cast him as a military magician, a warrior-scientist / one-man death squad. According to these legends, when his home city of Syracuse was under siege by the Romans in 212 BCE, Archimedes — by then an old man, around seventy — helped defend the city by using his knowledge of pulleys and levers to make fantastical weapons, "war engines" such as grappling hooks and giant cranes that could lift the Roman ships out of the sea and shake the sailors from them like sand being shaken out of a shoe. As Plutarch described the terrifying scene, "A ship was frequently lifted up to a great height in the air (a dreadful thing to behold), and was rolled to and fro, and kept swinging, until the mariners were all thrown out, when at length it was dashed against the rocks, or let fall."

In a more serious vein, all students of science and engineering remember Archimedes for his principle of buoyancy (a body immersed in a fluid is buoyed up by a force equal to the weight of the fluid displaced) and his law of the lever (heavy objects placed on opposite sides of a lever will balance if and only if their weights are in inverse proportion to their distances from the fulcrum). Both of these ideas have countless practical applications. Archimedes's principle of buoyancy explains why some objects float and others do not. It also underlies all of naval architecture, the theory of ship stability, and the design of oil-drilling platforms at sea. And you rely on his law of the lever, even if unknowingly, every time you use a nail clipper or a crowbar.

Archimedes might have been a formidable maker of war machines, and he undoubtedly was a brilliant scientist and engineer, but what really puts him in the pantheon is what he did for mathematics. He paved the way for integral calculus. Its deepest ideas are plainly visible in his work, but then they aren't seen again for almost two millennia. To say he was ahead of his time would be putting it mildly. Has anyone ever been *more* ahead of his time?

Two strategies appear again and again in his work. The first was his ardent use of the Infinity Principle. To probe the mysteries of circles, spheres, and other curved shapes, he always approximated them with rectilinear shapes made of lots of straight, flat pieces, faceted like jewels. By imagining more and more pieces and making them smaller and smaller, he pushed his approximations ever closer to the truth, approaching exactitude in the limit of infinitely many pieces. This strategy demanded that he be a wizard with sums and puzzles, since he ended up having to add many numbers or pieces back together to arrive at his conclusions.

His other distinguishing stratagem was blending mathematics with physics, the ideal with the real. Specifically, he mingled geometry, the study of shapes, with mechanics, the study of motion and force. Sometimes he used geometry to illuminate mechanics; sometimes the flow went in the other direction, with mechanical arguments providing insight into pure form. It was by using both strategies with consummate skill that Archimedes was able to penetrate so deeply into the mystery of curves.

Squeezing Pi

When I walk to my office or go out with my dog for an evening stroll, the pedometer on my iPhone keeps track of how far I walk. The calculation is simple: The app estimates the length of my stride based on my height and counts how many steps I've taken, then it multiplies those two numbers together. The distance traveled equals stride length times the number of steps taken.

Archimedes used a similar idea to calculate the circumference of a circle and to estimate pi. Think of the circle as a track. It takes a lot of steps to walk all the way around. The path would look something like this.

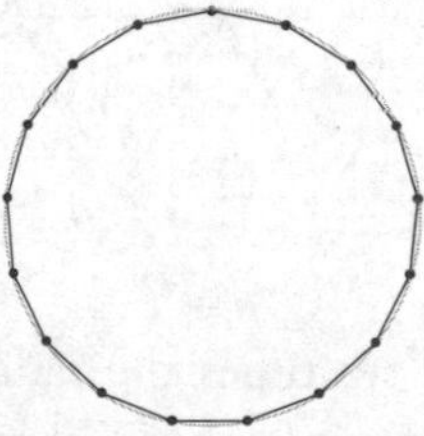

Each step is represented by a tiny straight line. By multiplying the number of steps by the length of each one, we can estimate the length of the track. It's only an estimate, of course, because the circle is not actually made up of straight lines. It's made up of curved arcs. When we replace each arc by a straight line, we're taking a slight shortcut. And so the approximation is sure to *underestimate* the true length of the circular track. But, at least in theory, by taking enough steps and making them small enough, we can approximate the length of the track as accurately as we wish.

Archimedes did a series of calculations like this, starting with a path made up of six straight steps.

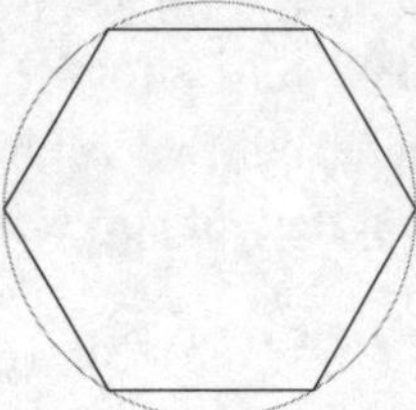

He began with a hexagon because it was a convenient base camp from which to embark on the more arduous calculations ahead. The advantage of the hexagon was that he could easily calculate its perimeter, the total length around the hexagon. It's six times the radius of the circle. Why six? Because the hexagon contains six equilateral triangles, each side of which equals the circle's radius.

Six of the triangle's sides make up the perimeter of the hexagon.

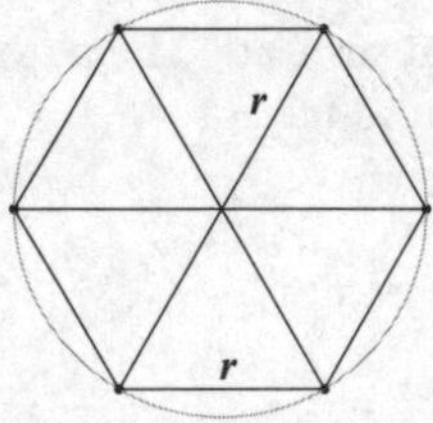

So the perimeter equals six times the radius; in symbols, $p = 6r$. Then, since the circle's circumference C is longer than the hexagon's perimeter p, we must have $C > 6r$.

This argument gave Archimedes a lower bound on what we would call pi, written as the Greek letter π and defined as the ratio of the circumference to the diameter of the circle. Since the diameter d equals $2r$, the inequality $C > 6r$ implies

$$\pi = \frac{C}{d} = \frac{C}{2r} > \frac{6r}{2r} = 3.$$

Thus the hexagon argument demonstrates $\pi > 3$.

Of course, six is a ridiculously small number of steps, and the resulting hexagon is obviously a very crude caricature of a circle, but Archimedes was just getting started. Once he figured out what the hexagon was telling him, he shortened the steps and took twice as many of them. He did that by detouring to the midpoint of each arc, taking two baby steps instead of striding across the arc in one big step.

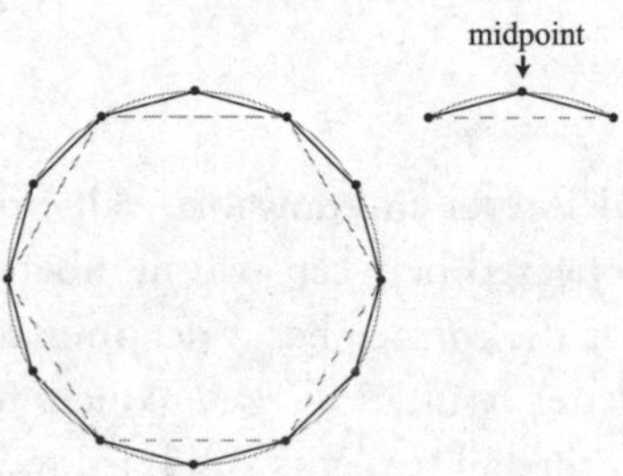

Then he kept doing that, over and over again. A man obsessed, he went from six steps to twelve, then twenty-four, forty-eight, and, ultimately, ninety-six steps, working out their ever-shrinking lengths to migraine-inducing precision.

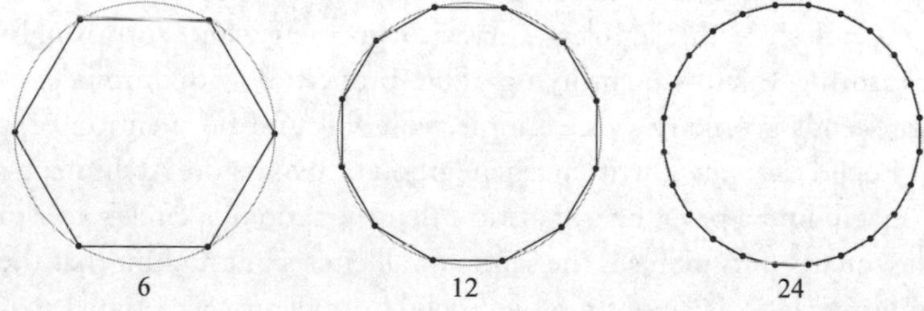

Unfortunately, it got progressively harder to calculate the step lengths as they shrank, because he had to keep invoking the Pythagorean theorem to find them. That required him to calculate square roots, a nasty chore to do by hand. Furthermore, to ensure that he was always underestimating the circumference, he had to make sure that his approximating fractions bounded the bothersome square roots from below when he needed them to be underestimates and from above when he needed them to be overestimates.

What I'm trying to say is that his calculation of π was heroic, both logically and arithmetically. By using a 96-gon inside the circle and a 96-gon outside the circle, he ultimately proved that π is greater than $3 + 10/71$ and less than $3 + 10/70$.

Forget about math for a minute. Just savor this result at a visual level:

$$3+\tfrac{10}{71}<\pi<3+\tfrac{10}{70}\,.$$

The unknown, and forever unknowable, value of π is trapped in a numerical vise, squeezed between two numbers that look almost identical except that the former has a denominator of 71 and the latter of 70. That latter result, $3 + 10/70$, reduces to $22/7$, the famous approximation to π that all students still learn today and that some unfortunately mistake for π itself.

The squeeze technique that Archimedes used (building on earlier work by the Greek mathematician Eudoxus) is now known as the method of exhaustion because of the way it traps the unknown number pi between two known numbers. The bounds tighten with each doubling, thus exhausting the wiggle room for pi.

Circles are the simplest curves in geometry. Yet, surprisingly, measuring them—quantifying their properties with numbers—transcends geometry. For example, you will find no mention of π in Euclid's *Elements*, written a generation or two before Archimedes. You will find a proof by exhaustion that the ratio of a circle's area to the square of its radius is the same for all circles but no hint that the universal ratio is close to 3.14. Euclid's omission was a signal that something deeper was needed. To come to grips with π's numerical value required a new kind of mathematics, one that could cope with curved shapes. How to measure the length of a curved line or the area of a curved surface or the volume of a curved solid—these were the cutting-edge questions that consumed Archimedes and led him to take the first steps toward what we now call integral calculus. Pi was its first triumph.

The Tao of Pi

It may seem strange to modern minds that pi doesn't appear in Archimedes's formula for the area of a circle, $A = rC/2$, and that he never wrote down an equation like $C = \pi d$ to relate the circumference of a circle to its diameter. He avoided doing all that because pi was not a number to him. It was simply a ratio of two lengths, a proportion between a circle's circumference and its diameter. It was a magnitude, not a number.

We no longer make this distinction between magnitude and number, but it was important in ancient Greek mathematics. It seems to have arisen from the tension between the discrete (as represented by whole numbers) and the continuous (as represented by shapes). The historical details are murky, but it appears that sometime between Pythagoras and Eudoxus, between the sixth and the fourth centuries BCE, somebody proved that the diagonal of a square was incommensurable with its side, meaning that the ratio of those two lengths could not be expressed as the ratio of two whole numbers. In modern language, someone discovered the existence of irrational numbers. The suspicion is that this discovery shocked and disappointed the Greeks, since it belied the Pythagorean credo. If whole numbers and their ratios couldn't even measure something as basic as the diagonal of a perfect square, then all was *not* number. This deflating letdown may explain why later Greek mathematicians always elevated geometry over arithmetic. Numbers couldn't be trusted anymore. They were inadequate as a foundation for mathematics.

To describe continuous quantities and reason about them, the ancient Greek mathematicians realized they needed to invent something more powerful than whole numbers. So they developed a system based on shapes and their proportions. It relied on measures of geometrical objects: lengths of lines, areas of squares, volumes of cubes. All of these they called magnitudes. They thought of them as distinct from numbers and superior to them.

This, I believe, is why Archimedes held pi at arm's length. He didn't know what to make of it. It was a strange, transcendent creature, more exotic than any number.

Today we accept pi as a number—a real number, an infinite decimal—and a fascinating one at that. My children certainly were intrigued by it. They used to stare at a pie plate hanging in our kitchen that had the digits of pi running around the rim and spiraling in toward the center, shrinking in size as they swirled into the abyss. For them, the fascination had to do with the random-looking sequence of digits, never repeating, never showing any pattern at all, going on forever, infinity on a platter. The first few digits in pi's infinite decimal expansion are

3.14159265358979323846264338327950288419716939
93751058209749 . . .

We will never know all the digits of pi. Nevertheless, those digits are out there, waiting to be discovered. As of this writing, twenty-two trillion digits have been computed by the world's fastest computers. Yet twenty-two trillion is nothing compared to the infinitude of digits that define the actual pi. Think of how philosophically disturbing this is. I said that the digits of pi are out there, but where are they exactly? They don't exist in the material world. They exist in some Platonic realm, along with abstract concepts like truth and justice.

There's something so paradoxical about pi. On the one hand, it represents order, as embodied by the shape of a circle, long held to be a symbol of perfection and eternity. On the other hand, pi is unruly, disheveled in appearance, its digits obeying no obvious rule, or at least none that we can perceive. Pi is elusive and mysterious, forever beyond reach. Its mix of order and disorder is what makes it so bewitching.

Pi is fundamentally a child of calculus. It is defined as the unattainable limit of a never-ending process. But unlike a sequence of polygons steadfastly approaching a circle or a hapless walker stepping halfway to a wall, there is no end in sight for pi, no limit we can ever know. And yet pi exists. There it is, defined so crisply as the ratio of two lengths we can see right before us, the circumference of a circle and its diameter. That ratio defines pi, pinpoints it as clearly as can be, and yet the number itself slips through our fingers.

With its yin and yang binaries, pi is like all of calculus in miniature. Pi is a portal between the round and the straight, a single number yet infinitely complex, a balance of order and chaos. Calculus, for its part, uses the infinite to study the finite, the unlimited to study the limited, and the straight to study the curved. The Infinity Principle is the key to unlocking the mystery of curves, and it arose here first, in the mystery of pi.

Cubism Meets Calculus

Archimedes went deeper into the mystery of curves, again guided by the Infinity Principle, in his treatise *The Quadrature of the Parabola.* A parabola describes the familiar arc of a three-point shot in basketball or water coming out of a drinking fountain. Actually, those arcs in the real world are only approximately parabolic. A true parabola, to Archimedes, would have meant a curve obtained by slicing through a cone with a plane. Imagine a meat cleaver slicing through a dunce cap or a conical paper cup; the cleaver can make different kinds of curves depending on how steeply it cuts through the cone. A slice parallel to the base of a cone makes a circle.

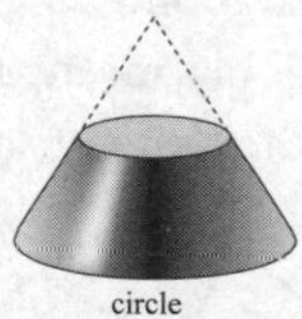

A slightly steeper cut produces an ellipse.

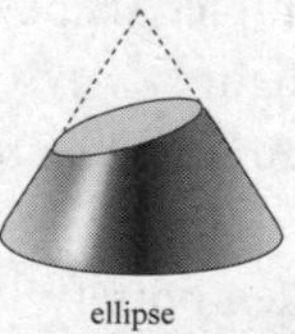

A cut that has the same slope as the cone itself produces a parabola.

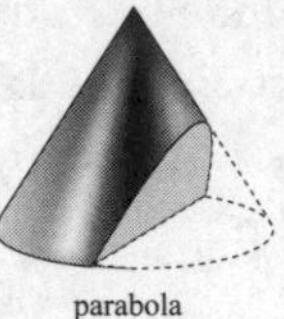

Viewed in the plane of the slice, the parabola appears as a graceful, symmetrical curve with a line of symmetry down its middle. This line is called the parabola's axis.

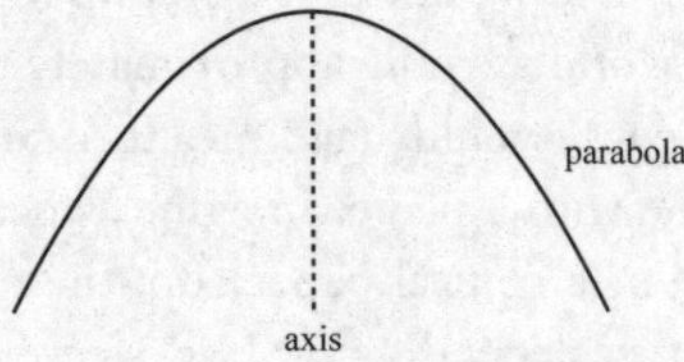

In his treatise, Archimedes set himself the challenge of working out the quadrature of a parabolic segment. In more modern language, a segment of a parabola means the curved region lying between the parabola and a line that cuts across it obliquely.

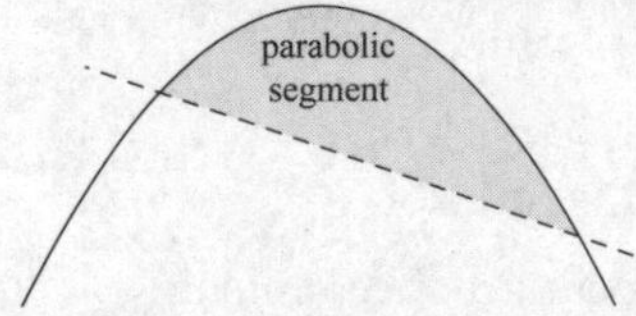

Finding its quadrature means expressing its unknown area in terms of the known area of a simpler shape like a square, rectangle, triangle, or other rectilinear figure.

The strategy used by Archimedes was astonishing. He reimagined the parabolic segment as infinitely many triangular shards glued together like pieces of broken pottery.

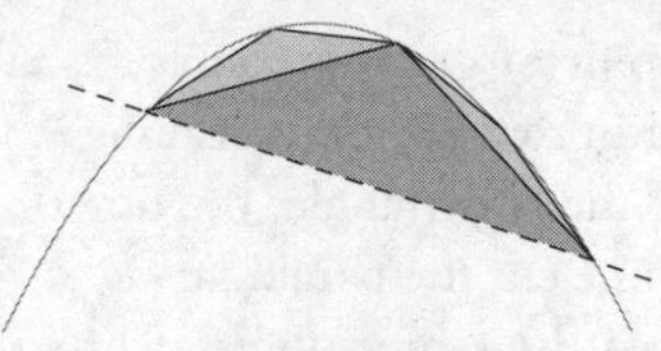

The shards came in an endless hierarchy of sizes: one big triangle, two smaller ones, four smaller still, and so on. His plan was to find all their areas and then add them back together to calculate the curved area he was wondering about. It took a kaleidoscopic leap of artistic imagination to see a smooth, gently curving parabolic segment as a mosaic of jagged shapes. If he had been a painter, Archimedes would have been the first cubist.

To carry out his strategy, Archimedes first had to find the areas of all the shards. But how, precisely, were those shards to be defined? After all, there are countless ways to piece triangles together to form a parabolic segment, just as there are countless ways to smash a plate into jagged bits. The biggest triangle could look like this, or this, or this:

He came up with a brilliant idea—brilliant because it established a rule, a consistent pattern that held from one level of the hierarchy to the next. He imagined sliding the oblique line at the base of the segment upward while keeping it parallel to itself until it just barely touched the parabola at a single point near the top.

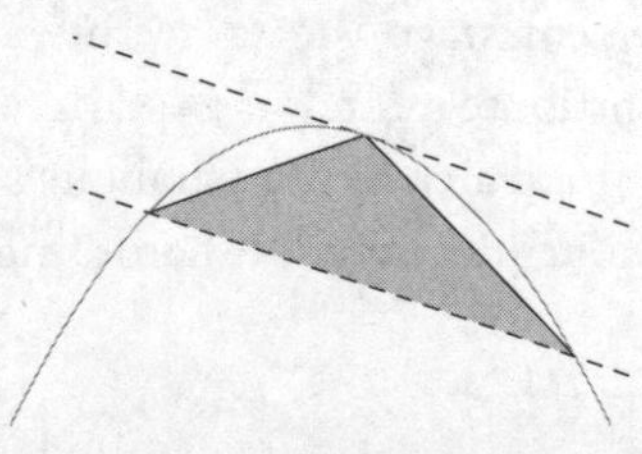

That special point of grazing contact is called a point of tangency (from the Latin root *tangere*, meaning "touching"). It defined the third corner of the big triangle, the other two being the points where the oblique line cut the parabola.

Archimedes used the same rule to define the triangles at *every* stage in the hierarchy. At the second stage, for example, the triangles looked like this.

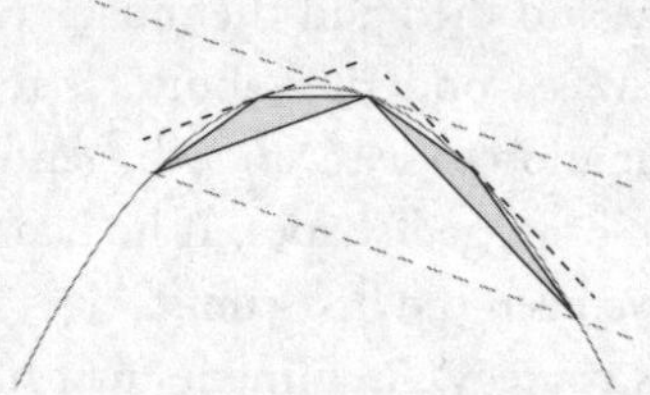

Notice that the sides of the big triangle now play the role of the oblique line used earlier.

Next, Archimedes invoked known geometrical facts about parabolas and triangles to relate one level of the hierarchy to the next. He proved that each newly created triangle had one-eighth as much area as its parent triangle. Thus, if we say that the first, biggest triangle occupies 1 unit of area—that triangle will serve as our area standard—then its two daughter triangles together occupy ⅛ + ⅛ = ¼ as much area.

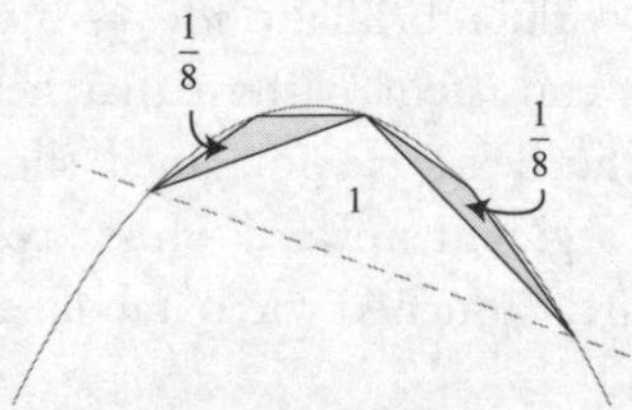

At each subsequent stage the same rule applies: the daughter triangles always contribute a total of a quarter as much area as their parent does. So the total area of the parabolic segment, reassembled from the whole infinite hierarchy of shards, must be

$$Area = 1 + \tfrac{1}{4} + \tfrac{1}{16} + \tfrac{1}{64} + \cdots,$$

an infinite series in which each term is one-quarter of the term preceding it.

There's a shortcut to sum this kind of infinite series, which is known in the trade as a geometric series. The trick is to cancel all but one of its infinitely many terms by multiplying both sides of the equation for *Area* by 4 and subtracting the original sum from it. Watch: Multiplying each term by 4 in the infinite series above gives

$$\begin{aligned} 4 \times Area &= 4\left(1+\tfrac{1}{4}+\tfrac{1}{16}+\tfrac{1}{64}+\cdots\right) \\ &= 4+\tfrac{4}{4}+\tfrac{4}{16}+\tfrac{4}{64}+\cdots \\ &= 4+1+\tfrac{1}{4}+\tfrac{1}{16}+\cdots \\ &= 4+Area. \end{aligned}$$

The magic happens between the next-to-last line and the last line above. The right-hand side of the last line equals $4 + Area$, because the original sum, $Area = 1+\tfrac{1}{4}+\tfrac{1}{16}+\cdots$, has, like a phoenix, been reborn in the terms following the 4 in the next-to-last line. So

$$4 \times Area = 4 + Area.$$

Subtract one *Area* from both sides to get $3 \times Area = 4$. Thus

$$Area = \tfrac{4}{3}.$$

In other words, the parabolic segment has $4/3$ the area of the big triangle.

A Cheesy Argument

Archimedes would not have approved of the legerdemain above. He arrived at the same result by a different route. He resorted to a subtle style of argumentation often described as double *reductio ad absurdum*, a double proof by contradiction. He proved that the area of the parabolic segment could not be less than $4/3$ or greater than $4/3$, so it must equal $4/3$. As Sherlock Holmes later put it, "When you have eliminated the impossible, whatever remains, *however improbable,* must be the truth."

What's conceptually crucial here is that Archimedes eliminated the impossible with arguments based on a *finite* number of shards. He showed that their combined area could be made as close to $4/3$ as desired, closer than any prescribed tolerance, simply by taking enough of them. He never had to summon infinity. So everything about his proof was ironclad. It still meets the highest standards of rigor today.

The gist of his argument becomes easy to understand if we put it in everyday terms. Suppose three people want to share four identical slices of cheese.

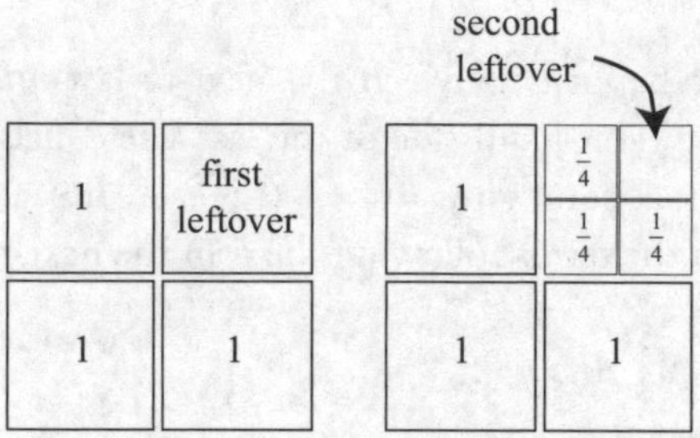

The commonsense solution would be to give each person a slice, then cut the remaining slice into thirds and hand them out. That's fair. In total, everyone would get $1 + 1/3 = 4/3$ of a slice.

But suppose the three people happen to be mathematicians who are milling around the food table before the seminar, eyeing the last four slices of cheese. The cleverest of the three, coincidentally named Archimedes, might suggest the following solution: "I'll take a slice and you guys take yours, which leaves one more for us to share. Euclid, cut that leftover slice into *quarters,* not thirds, and everyone, take a quarter of that leftover slice. We're going to keep doing this, always cutting what's left over into four equal portions, until the remaining crumb is of no interest to anyone. Okay? Eudoxus, stop whining."

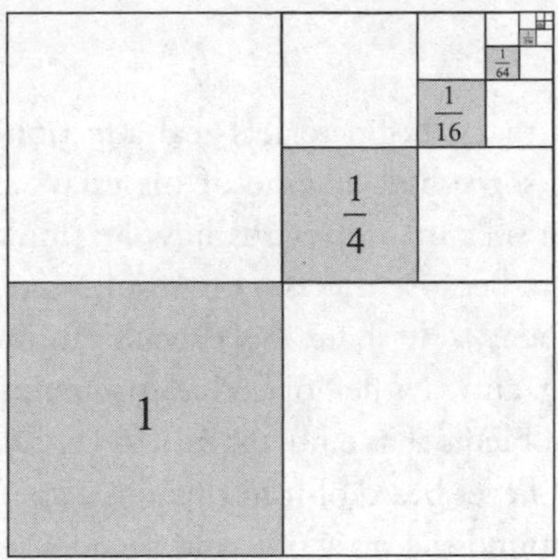

How many slices of cheese, total, would each of them get to eat if this were to go on indefinitely? One way to look at it is to keep a running tally of how many slices each person gets. After round one, each gets one slice. After round two, when the quarter slices are passed out, each person has accumulated $1 + \frac{1}{4}$ slices. After round three, when the quarters are themselves quartered into sixteenths, the running total for each is $1 + \frac{1}{4} + \frac{1}{16}$ slices. And so on. Loosely speaking, each of the three people would eventually get to eat $1 + \frac{1}{4} + \frac{1}{16} + \cdots$ slices in total if the cutting went on forever. And since this amount must represent a third of the original four slices, it must be that $1 + \frac{1}{4} + \frac{1}{16} + \cdots$ equals one-third of 4, which is $\frac{4}{3}$.

In *The Quadrature of the Parabola*, Archimedes gave an argument very close to this, including a diagram with squares of different sizes, but he never invoked infinity or used the counterpart of the three dots [$\cdots$] above to signify that the sum went on endlessly. Rather, he phrased his argument in terms of finite sums so that it was unimpeachably rigorous. His key observation was that the tiny square in the upper right corner—the current leftover remaining to be shared—could be made smaller than any given amount by considering a sufficiently large but finite number of rounds. And by similar reasoning, the finite sum $1 + \frac{1}{4} + \frac{1}{16} + \cdots + \frac{1}{4^n}$ (the total amount of cheese that each person gets) could be made as close to $\frac{4}{3}$ as desired by making n large enough. So the only possible answer was $\frac{4}{3}$.

The Method

It's at this point that I begin to feel real affection for Archimedes, because he does something in one of his essays that few geniuses ever do: He invites us in and reveals how he thinks. (I'm using the present tense here because the essay is so intimate, it feels like he's speaking to us today.) He shares his private intuition, a vulnerable, soft-bellied thing, and says he hopes that future mathematicians will use it to solve problems that eluded him. Today this secret is known as the Method. I never heard of it in calculus class. We don't teach it anymore. But I found the story of it and the idea behind it enthralling and astounding.

He writes about it in a letter to his friend Eratosthenes, the librarian at Alexandria and the only mathematician of his era who could understand him. He confesses that even though his Method "does not furnish an actual demonstration" of the results he's interested in, it helps him figure out what's true. It gives him intuition. As he says, "It is easier to supply the proof when we have previously acquired, by the method, some knowledge of the questions than it is to find it without any previous knowledge." In other words, by noodling around, playing with the Method, he gets a feel for the territory. And that guides him to a watertight proof.

This is such an honest account of what it's like to do creative mathematics. Mathematicians don't come up with the proofs first. First comes intuition. Rigor comes later. This essential role of intuition and imagination is often left out of high-school geometry courses, but it is essential to all creative mathematics.

Archimedes concludes with the hope that "there will be some among the present as well as future generations who by means of the method here explained will be enabled to find other theorems which have not yet fallen to our share." That almost brings a tear to my eye. This unsurpassed genius, feeling the finiteness of his life against the infinitude of mathematics, recognizes that there is so much left to be done, that there are "other theorems which have not yet fallen to our share." We all feel that, all of us mathematicians. Our subject is endless. It humbled even Archimedes himself.

The first mention of the Method appears at the beginning of the essay on the quadrature of the parabola, before the cubist proof with the shards. Archimedes confesses that the Method led him to that proof and to the number 4/3 in the first place.

What is the Method, and what is so personal, brilliant, and transgressive about it? The Method is *mechanical;* Archimedes finds the area of the parabolic segment by *weighing* it in his mind. He thinks of the curved parabolic region as a material object—I'm picturing it as a thin sheet of metal carefully trimmed into the desired parabolic shape—and then he places it at one end of an imaginary balance scale. Or, if you prefer, think of it as being seated at one end of an imaginary seesaw. Next he figures out how to counterbalance it against a shape he already knows how to weigh: a triangle. From this he deduces the area of the original parabolic segment.

It's an even more imaginative approach than the cubist/geometric/shards-and-triangles technique of his that we discussed earlier, because in this case, he's going to build the imaginary seesaw as part of the calculation and design it to comport with the parabola's dimensions. Together, they will produce the answer he seeks.

He starts with the parabolic segment and tilts it to ensure that the parabola's symmetry axis is vertical.

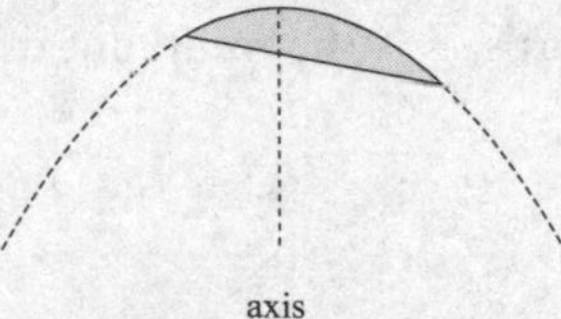

Then he builds the seesaw around it. The instruction manual reads as follows: *Draw the big triangle inside the parabolic segment, as before, and label it ABC.* As in the cubist proof, this triangle is again going to serve as an area standard. The parabolic segment will be compared to it and will turn out to have four-thirds its area.

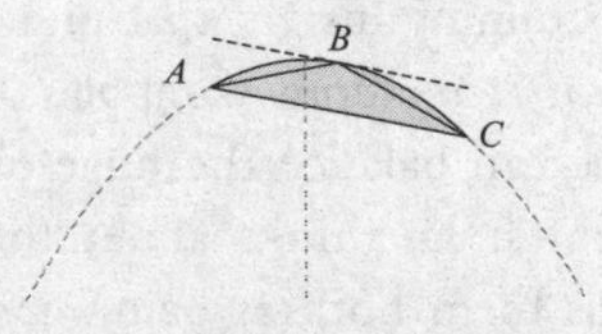

Next enclose the parabolic segment in a much bigger triangle, ACD.

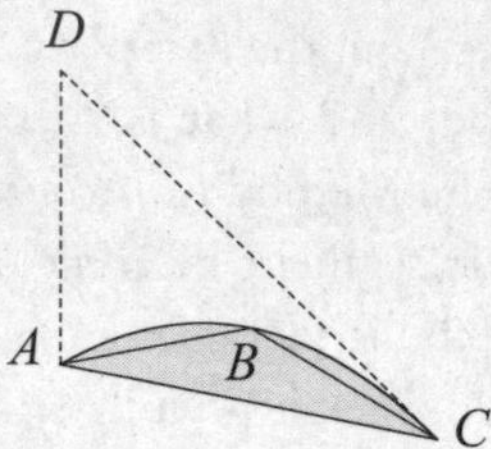

The triangle's top side is chosen to be a line tangent to the parabola at the point C. Its base is the line AC. And its left side is a vertical line that extends upward from A until it meets the top side at point D. Using standard Euclidean geometry, Archimedes proves that this huge outer triangle ACD has four times the area of the inner triangle ABC. (That fact will become important later. Set it aside for now.)

The next step is to build the rest of the seesaw—its lever, its two seats, and its fulcrum. The lever is the line that joins the two seats. That line starts at C, goes through B, emerges from the huge outer triangle at F (the fulcrum), and continues to the left until it hits a point S (the seat). The condition that defines S is that it's as far from F as C is. In other words, F is the midpoint of the line SC.

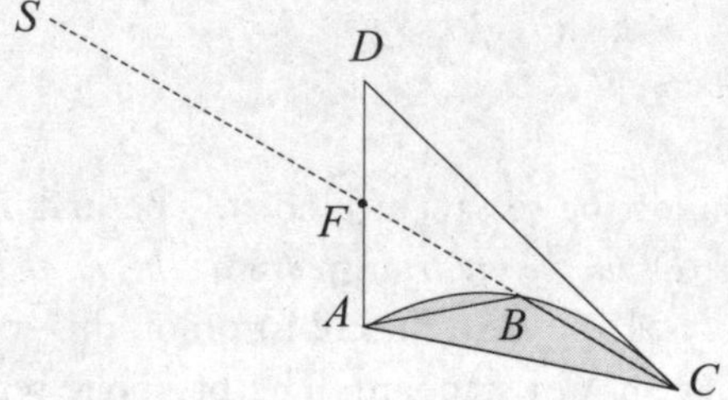

Now comes the stunning insight that underlies the whole conception. Using known facts about parabolas and triangles, Archimedes proves that he can balance the huge outer triangle against the parabolic segment if he thinks about them *one vertical line at a time*. He regards them both as being composed of infinitely many parallel lines. Those lines are like infinitesimally thin slats or ribs. Here's a typical pair of them, defined by a single vertical line through both shapes. On that line, a short rib connects the base to the parabola,

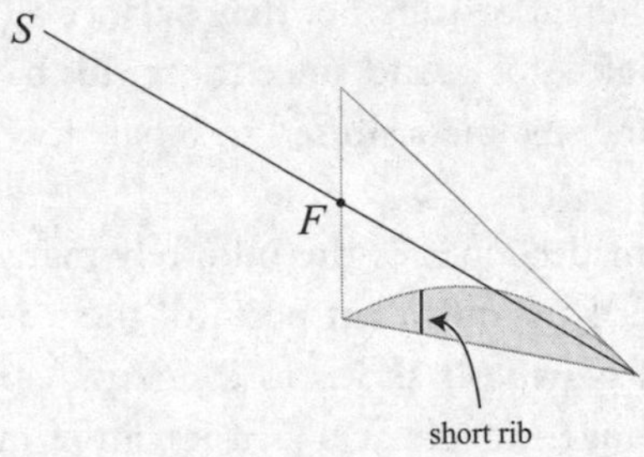

and a tall rib connects the base to the top side of the huge outer triangle.

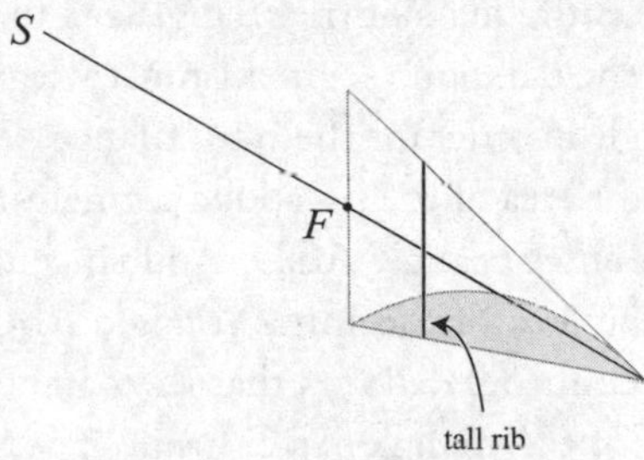

His amazing insight is that these ribs balance each other perfectly, like kids playing on a seesaw, as long as they sit in the right places. He proves that if he slides the short rib over to the point S and leaves the tall rib in place, they balance.

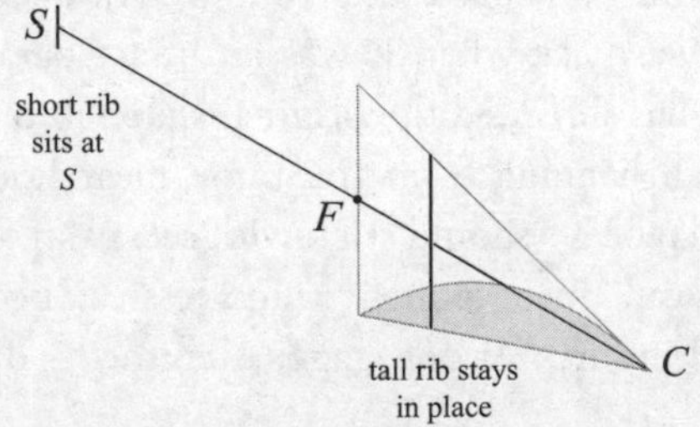

The same is true for *every* vertical slice. No matter which vertical slice you take, the short rib always balances the tall rib if you slide the short rib to S and leave the tall rib in place.

So the two shapes balance each other, rib by rib. All the ribs from the parabola end up at S. Together they balance all the ribs from the huge outer triangle ACD. And since those ribs haven't moved, that means all the parabolic mass shifted to S balances the huge triangle right where it is.

Next, Archimedes replaces the infinitely many ribs of the huge outer triangle with an equivalent point of their own, called the triangle's center of gravity. It serves as a proxy. As far as seesaws are concerned, the huge triangle acts as if its entire mass were concentrated at that single center of gravity. That location, Archimedes has already shown in other work, lies on the line FC at a point precisely three times closer to the fulcrum F than S is.

So, since the entire mass of the triangle sits three times closer to the pivot point, the parabolic segment must weigh a third as much as the huge triangle in order for them to balance; that's the law of the lever. Therefore the area of the parabolic segment must be one-third that of the huge outer triangle ACD. And since that outer triangle has four times the area of the inner triangle ABC (the fact we set aside earlier), Archimedes deduces that the parabolic segment must have 4⁄3 the area of the triangle ABC inside it . . . just as we found earlier by summing the infinite series of triangular shards!

I hope I've managed to convey what an acid trip of an argument this is. Instead of a potter reassembling shards, here Archimedes is more like a butcher. He takes the tissue of the parabolic region apart, one vertical strip at a time, and hangs all these infinitesimally thin strips of flesh from a hook at S. The total weight of all the flesh stays the same as it was back when it was an intact parabolic segment. It's just that he has shredded the original shape into lots of vertical, stringy strips, all hanging from the same meat hook. (It's such a weird image. Maybe we should stick with seesaws.)

Why did I call this argument transgressive? Because it traffics with completed infinity. At one stage, Archimedes openly describes the outer triangle as being "made up of all the parallel lines" inside itself. That, of course, is taboo in Greek mathematics; there's a continuous infinity of these parallel lines, these vertical ribs. He's openly thinking of the triangle as a completed infinity of ribs. In doing so, he's unleashing the golem.

Likewise he describes the parabolic segment as being "made up of all the parallel lines drawn inside the curve." Dallying with completed infinity lowers the status of this reasoning, in his estimation, to a heuristic—a means of finding an answer, not a proof of its correctness. In his letter to Eratosthenes, he downplays the Method as giving nothing more than "a sort of indication" that the conclusion is true.

Whatever its logical status, Archimedes's Method has an *e pluribus unum* quality to it. This Latin phrase, the motto of the United States, means "out of many, one." Out of the infinitely many straight lines making up the parabola, one area emerges. Thinking of that area as a mass, Archimedes shifts it, line by line, to the far left seat on the seesaw. The infinitude of lines is thereby represented by a single mass seated at a single point. The one replaces the many and stands for it, representing it perfectly and faithfully.

The same is true for the counterbalancing outer triangle on the right of the seesaw. Out of its continuum of vertical lines, one point is chosen—its center of gravity. It too stands for the whole. Infinity collapses to unity; *e pluribus unum*. Except this is not poetry or politics. This is the beginning of integral calculus. Triangles and parabolic regions are apparently and mysteriously equivalent, in some sense that Archimedes could not quite make rigorous, to infinitudes of vertical lines.

Although Archimedes seems embarrassed by his dalliance with infinity, he is brave enough to own up to it. Anyone trying to measure a curved shape—to find the length of its boundary or the area or the volume inside it—has to grapple with the limit of an infinite sum of infinitesimally small pieces. Careful souls may try to sidestep that necessity, finessing it with the method of exhaustion. But at bottom, there is no escaping it. Coping with curved shapes means coping with infinity, one way or another. Archimedes is open about this. When he needs to, he can dress up his proofs in respectable garb, sporting finite sums and the method of exhaustion. But in private, he's dirty. He admits to weighing shapes in his mind, dreaming of levers and centers of gravity, balancing regions and solids line by line, one infinitesimal piece at a time.

Archimedes went on to apply the Method to many other problems about curved shapes. For example, he used it to discover the center of gravity of a solid hemisphere, a paraboloid, and segments of ellipsoids and hyperboloids. His favorite result, which he loved so much that he asked that it be carved on his tombstone, concerned the surface area and volume of a sphere.

Picture a sphere sitting snugly in a cylindrical hatbox.

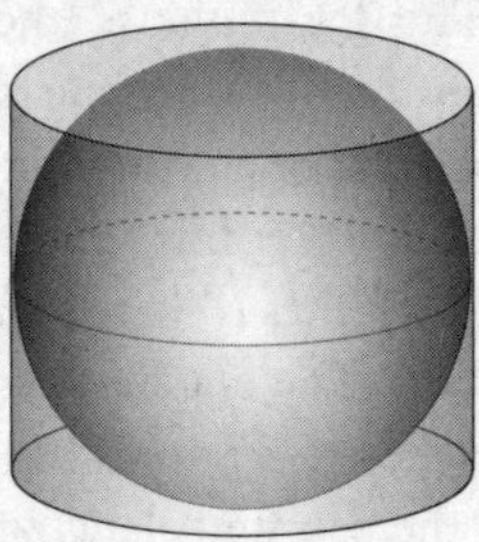

Using the Method, Archimedes discovered that the sphere has ⅔ the volume of the enclosing hatbox, as well as ⅔ of its surface area (assuming the top and bottom lids are also counted in the hatbox's surface area). Notice that he didn't give *formulas* for the volume or the surface area of the sphere, as we would today. Rather, he phrased his results as proportions. That's classic Greek style. Everything was expressed as a proportion. An area was compared to another area, a volume to another volume. And when their ratio involved small whole numbers, as they do here with 3 and 2 and as they did with 4 and 3 in the quadrature of the parabola, that must have been a source of particular pleasure to him. After all, those same ratios, 3:2 and 4:3, held special significance to the ancient Greeks because of their central role in the Pythagorean theory of musical harmony.

Recall that when two otherwise identical strings with lengths in the ratio 3:2 are struck, they harmonize beautifully, separated in pitch by an interval known as a fifth. Similarly, strings in a 4:3 ratio produce a fourth. These numerical coincidences between harmony and geometry must have delighted Archimedes.

His words in his essay "On the Sphere and Cylinder" suggest just how tickled he was: "Now these properties were all along naturally inherent in the figures, but remained unknown to those who were before my time engaged in the study of geometry." Ignore how proud he sounds and focus instead on his claim that the properties he discovered "were all along naturally inherent in the figures, but remained unknown." Here he is expressing a particular philosophy of mathematics dear to the hearts of all working mathematicians. We feel we are *discovering* mathematics. The results are there, waiting for us. They have been inherent in the figures all along. We are not inventing them. Unlike Bob Dylan or Toni Morrison, we are not creating music or novels that never existed before; we are discovering facts that already exist, that are inherent in the objects we study. Although we have creative freedom to invent the objects themselves — to create idealizations like perfect spheres and circles and cylinders — once we do, they take on lives of their own.

When I read the way Archimedes expresses his pleasure at unveiling the surface area and volume of the sphere, I feel like I'm feeling the same things he felt. Or, rather, that *he* was feeling the same things I feel and that *all* of my colleagues feel when we do mathematics. Although we are told that the past is a foreign country, it may not be foreign in every respect. People we read about in Homer and the Bible seem a lot like us. And the same appears to be true of ancient mathematicians, or at least of Archimedes, the only one who let us into his heart.

Twenty-two centuries ago, Archimedes wrote a letter to his friend Eratosthenes, the librarian at Alexandria, essentially sending him a mathematical message in a bottle that virtually no one could appreciate but that he hoped might somehow sail safely across the seas of time. He had shared his private intuition, his Method, in the wish that it might enable future generations of mathematicians "to find other theorems which have not yet fallen to our share." The odds were against him. As always, the ravages of time were cruel. Kingdoms fell and libraries were burned. Manuscripts decayed. Not a single copy of the Method was known to have survived the Middle Ages. Although Leonardo da Vinci, Galileo, Newton, and other geniuses of the Renaissance and the scientific revolution pored over what was left of Archimedes's treatises, they never had a chance to read the Method. It was thought to be irretrievably lost.

And then, miraculously, it was found.

In October 1998 a battered medieval prayer book came up for auction at Christie's and sold to an anonymous private collector for $2.2 million. Barely visible under its Latin prayers lay faint geometrical diagrams and mathematical text written in tenth-century Greek. The book is a palimpsest; in the thirteenth century, its parchment folios had been washed and scraped clean of the original Greek and overwritten with Latin liturgical text. Fortunately, the Greek was not completely obliterated. It contains the only surviving copy of Archimedes's Method.

The Archimedes Palimpsest, as it is now known, first came to light in 1899 in a Greek Orthodox library in Constantinople. It spent the Renaissance and the scientific revolution undetected in a prayer book in the monastery of St. Sabas near Bethlehem. It now lives in the Walters Art Museum in Baltimore, where it has been lovingly restored and examined using the latest imaging technology.

Archimedes Today: From Computer Animation to Facial Surgery

Archimedes's legacy lives on today. Consider the computer-animated movies that our kids love to watch. The characters in films like *Shrek*, *Finding Nemo*, and *Toy Story* seem lifelike and real, in part because they embody an Archimedean insight: Any smooth surface can be convincingly approximated by triangles. For example, here are three triangulations of a mannequin's head.

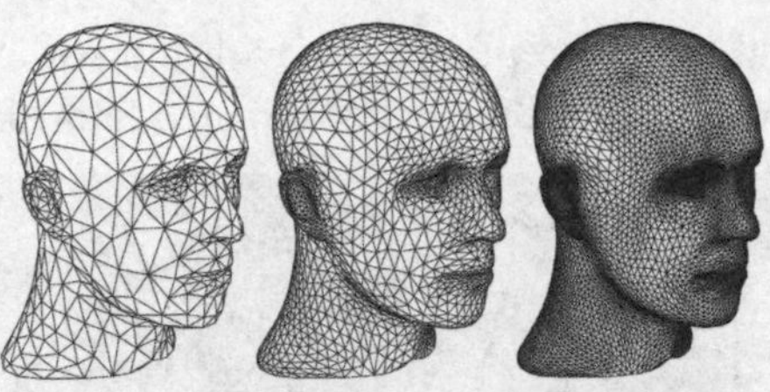

The more triangles we use and the smaller we make them, the better the approximation becomes.

What's true for mannequins is equally true for ogres, clownfish, and toy cowboys. Just as Archimedes used a mosaic of infinitely many triangular shards to represent a segment of a smoothly curved parabola, modern-day animators at DreamWorks created Shrek's round belly and his cute little trumpet-like ears out of tens of thousands of polygons. Even more were required for a tournament scene in which Shrek battled local thugs; each frame of that scene took over forty-five million polygons. But there was no trace of them anywhere in the finished movie. As the Infinity Principle teaches us, the straight and the jagged can impersonate the curved and the smooth.

When *Avatar* was released nearly a decade later, in 2009, the level of polygonal detail became more extravagant. At director James Cameron's insistence, animators used about a million polygons to render each plant on the imaginary world of Pandora. Given that the movie took place in a lush virtual jungle, that amounted to a *lot* of plants . . . and a lot of polygons. No wonder *Avatar* cost three hundred million dollars to produce. It was the first movie to use polygons by the billions.

The earliest computer-generated movies used far fewer polygons. Nonetheless, the computations seemed staggering at the time. Consider *Toy Story,* released in 1995. Back then, it took a single animator a week to sync an eight-second shot. The whole film took four years and eight hundred thousand hours of computer time to complete. As Pixar co-founder Steve Jobs told *Wired,* "There are more PhDs working on this film than any other in movie history."

Soon after *Toy Story* came *Geri's Game,* the first computer-animated film with a human main character. This funny/sad story of a lonesome old man who plays chess with himself in the park won the 1998 Academy Award for Best Animated Short Film.

Like other characters generated by a computer, Geri was built from angular shapes. At the beginning of this section, I showed a computer graphic of a face made from ever more triangles. In much the same way, the animators at Pixar fashioned Geri's head from a complex polyhedron, a three-dimensional gem-like shape that consisted of about forty-five hundred corners with flat facets in between them. The animators subdivided those facets repeatedly to create an increasingly detailed depiction. This subdivision process took up much less memory in the computer than earlier methods had, and it allowed for much faster animations. It was a revolutionary advance in computer animation at the time. But in spirit, it channeled Archimedes. Recall that to estimate pi, Archimedes started with a hexagon, then subdivided each of its sides and pushed their midpoints out to the circle to generate a 12-gon. After another subdivision, the 12-gon became a 24-gon, then a 48-gon, and finally a 96-gon, each encroaching ever more closely on its target, a limiting circle. Likewise, Geri's animators approximated the character's wrinkly forehead, his protuberant nose, and the folds of skin in his neck by repeatedly subdividing a polyhedron. By repeating that process enough times, they could make Geri look like what he was intended to be, a puppet-like character who conveyed a wide range of human feeling.

A few years later, a Pixar rival, DreamWorks, took the next steps forward in realism and emotional expressiveness in their story of a smelly, grouchy, heroic ogre named Shrek.

Although he never existed outside a computer, Shrek seemed practically human. That was partly because the animators took such great care to reproduce human anatomy. Underneath his virtual skin, they built virtual muscle, fat, bones, and joints. It was done so faithfully that when Shrek opened his mouth to speak, the skin on his neck formed a double chin.

Which brings us to another field where Archimedes's idea of polygonal approximation has proved useful: facial surgery for patients with severe overbites, misaligned jaws, or other congenital malformations. In 2006, the German applied mathematicians Peter Deuflhard, Martin Weiser, and Stefan Zachow reported the results of their work using calculus and computer modeling to predict the outcomes of complex facial surgeries.

The team's first step was to build an accurate map of a patient's facial-bone structure. To do so, they scanned the patients with computerized tomography (CT) or magnetic resonance imaging (MRI). The results gave information about the three-dimensional configuration of facial bones in the skull, from which the researchers created a computer model of the patient's face. The model was not just geometrically accurate; it was biomechanically accurate. It incorporated realistic estimates of the material properties of skin and soft tissues such as fat, muscle, tendons, ligaments, and blood vessels. With the help of the computer model, surgeons could then perform operations on virtual patients, similar to how fighter pilots sharpen their skills in flight simulators. Virtual bones in the face, jaw, and skull could be cut, relocated, augmented, or removed entirely. The computer calculated how the virtual soft tissue behind the face would move and reconfigure itself in response to stresses produced by the face's new bone structure.

The results of such simulations were helpful in several ways. They alerted the surgeons to possible adverse effects the procedures could have on vulnerable structures like nerves, blood vessels, and the roots of teeth. They also revealed what the patient's face would look like postoperatively, since the model predicted how the soft tissues would reposition themselves after the patient healed. Another advantage was that the surgeons could prepare better for the actual operations in light of the simulated results. And the patients could make better decisions about whether to have the operations.

Archimedes came in when the researchers modeled the smooth two-dimensional surface of the skull with an enormous number of triangles. The soft tissue posed its own geometrical challenges. Unlike the skull, soft tissue forms a fully three-dimensional volume. It fills the complicated space in front of the skull and behind the skin of the face. The team represented it by hundreds of thousands of tetrahedrons, the three-dimensional counterparts of triangles. In the image below, the skull surface is approximated by 250,000 triangles (they're too small to be seen) and the volume of soft tissue consists of 650,000 tetrahedrons.

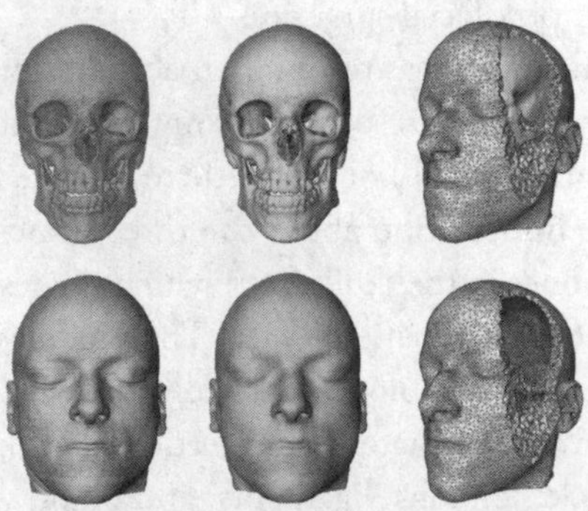

The array of tetrahedrons allowed the researchers to predict how the patient's soft tissues would deform after surgery. Roughly speaking, soft tissue is a deformable yet springy material, a bit like rubber or spandex. If you pinch your cheek, it changes shape; when you let go, it returns to normal. Ever since the 1800s, mathematicians and engineers have used calculus to model how different materials stretch, bend, and twist when they are pushed, pulled, or sheared in various ways. The theory is most highly developed in the more traditional parts of engineering, where it's used to analyze the stresses and strains in bridges, buildings, airplane wings, and many other structures made of steel, concrete, aluminum, and other hard materials. The German researchers adapted the traditional approach to soft tissues and found that it worked well enough to be valuable to surgeons and patients alike.

Their basic idea was this. Think of the soft tissue as a meshwork of tetrahedrons connected to one another like beads connected by elastic threads. The beads represent tiny portions of tissue. They are tied together elastically because, in reality, atoms and molecules in the tissue are linked by chemical bonds. Those bonds resist stretching and compression, which is what endows them with elasticity. During a virtual operation, a surgeon cuts bones in the virtual face and relocates some of the bone segments. When a piece of bone is moved to a new place, it pulls on the tissues it's connected to, which in turn pull on their neighboring tissues. The meshwork reconfigures itself due to the effect of cascading forces. As pieces of tissue move, they change the forces they exert on their neighbors by stretching or compressing the bonds between them. Those affected neighbors themselves readjust, and so on. Keeping track of all the resulting forces and displacements is a massive calculation that can be done only by computer. Step by step, an algorithm updates the myriad of forces and moves the tiny tetrahedrons accordingly. Ultimately all the forces balance and the tissue settles into its new equilibrium state. That's the new shape of the patient's face that the model predicts.

In 2006, Deuflhard, Weiser, and Zachow tested their model's predictions against the clinical outcomes of about thirty surgical cases. They found that the model worked remarkably well. As one measure of its success, it correctly forecast—to within one millimeter—the position of 70 percent of the patient's facial skin. Only 5 to 10 percent of the skin surface deviated by more than three millimeters from its predicted postoperative location. In other words, the model could be trusted. And it was certainly better than guesswork.

Here's an example of one patient before and after surgery. The four panels show his profile before the operation (far left), the computer model of his face at that time (mid-left), the predicted outcome of the surgery (mid-right), and the actual outcome (far right). Look at the position of his jaw before and after. The results speak for themselves.

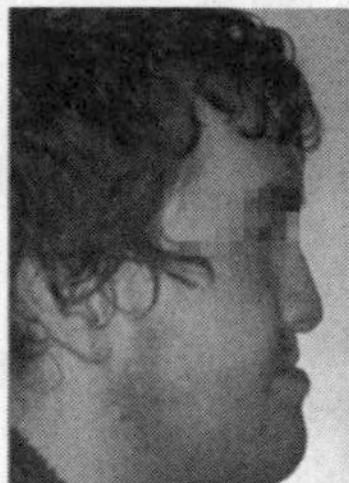  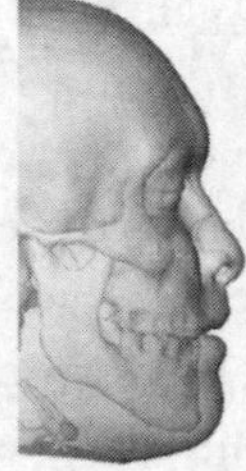 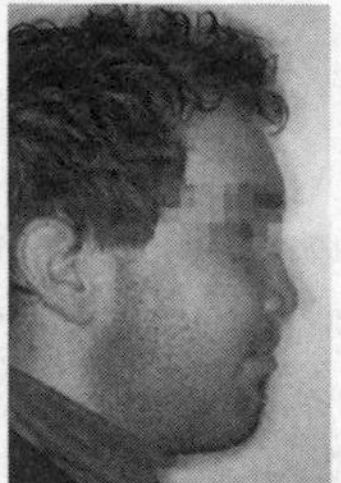

Onward to the Mystery of Motion

I am writing these words the day after a blizzard. Yesterday was March 14, Pi Day, and we got over a foot of snow. This morning, while I was shoveling my driveway for the fourth time, I watched jealously as a small tractor with a front-mounted snow thrower made its way easily down the sidewalk across the street. It used a rotating screw blade to pull snow into the machine and then ejected it onto my neighbor's yard.

This use of a rotating screw for propelling something goes back to Archimedes, at least according to legend. In his honor, today we call it an Archimedean screw. He is said to have come up with the invention during a trip to Egypt (although it may have been used much earlier by the Assyrians); it was developed to lift water from a low-lying area up to an irrigation ditch. Today, cardiac-assist devices use Archimedean-screw pumps to support circulation when the heart's left ventricle is impaired.

But apparently, Archimedes did not want to be remembered for his screws or his war engines or any other practical inventions; he never left us any writings about them. He was proudest of his inventions in mathematics. Which also gets me thinking that it is fitting to be reflecting on his legacy on Pi Day. In the twenty-two hundred years since Archimedes trapped pi, numerical approximations to pi have been improved many times, but always by using mathematical techniques that Archimedes himself introduced: approximations by polygons or by infinite series. More broadly, his legacy was the first principled use of infinite processes to quantify the geometry of curved shapes. At this he was unrivaled, and he remains so to this day.

Yet the geometry of curved shapes takes us only so far. We also need to know how things move in this world—how human tissue shifts after surgery, how blood flows through an artery, how a ball flies through the air. On this, Archimedes was silent. He gave us the science of statics, of bodies balancing on levers and floating stably in water. He was a master of equilibrium. The territory ahead concerned the mysteries of motion.

3

Discovering the Laws of Motion

When archimedes died, the mathematical study of nature nearly died along with him. Eighteen hundred years passed before a new Archimedes appeared. In Renaissance Italy, a young mathematician named Galileo Galilei picked up where Archimedes had left off. He watched how things moved when they flew through the air or fell to the ground, and he looked for numerical rules in their movements. He did careful experiments and made clever analyses. He timed pendulums swinging back and forth and rolled balls down gentle ramps and found marvelous regularities in both. Meanwhile, a young German mathematician named Johannes Kepler studied how the planets wandered across the sky. Both men were fascinated by patterns in their data and sensed the presence of something far deeper. They knew they were onto something, but couldn't quite make out its meaning. The laws of motion they were discovering were written in an alien language. That language, as yet unknown, was differential calculus. These were humanity's first hints of it.

Before the work of Galileo and Kepler, natural phenomena had rarely been understood in mathematical terms. Archimedes had revealed the mathematical principles of balance and buoyancy in his laws of the lever and hydrostatic equilibrium, but those laws were limited to static, motionless situations. Galileo and Kepler ventured beyond the static world of Archimedes and explored how things moved. Their struggles to make sense of what they saw spurred the invention of a new kind of mathematics that could handle motion at a variable rate. It addressed the type of change that keeps changing, like a ball gaining speed as it rolls down a ramp or the planets speeding up as they move closer to the sun and slowing down as they recede from it.

In 1623, Galileo described the universe as "this grand book . . . which stands continually open to our gaze," but cautioned that "the book cannot be understood unless one first learns to comprehend the language and read the letters in which it is composed. It is written in the language of mathematics, and its characters are triangles, circles, and other geometric figures without which it is humanly impossible to understand a single word of it; without these, one wanders about in a dark labyrinth." Kepler expressed even greater reverence for geometry. He described it as "coeternal with the divine mind" and believed that it "supplied God with patterns for the creation of the world."

The challenge for Galileo, Kepler, and other like-minded mathematicians of the early seventeenth century was to take their beloved geometry, so well suited to a world at rest, and extend it to a world in flux. The problems they faced were more than mathematical; they had to overcome philosophical, scientific, and theological resistance as well.

The World According to Aristotle

Before the seventeenth century, motion and change were poorly understood. Not only were they difficult to study; they were considered downright distasteful. Plato had taught that the object of geometry was to gain "knowledge of what eternally exists, and not of what comes for a moment into existence, and then perishes." His philosophical contempt for the transitory returned on a grander scale in the cosmology of his most illustrious student, Aristotle.

According to Aristotelian teaching, which dominated Western thought for almost two millennia (and which Catholicism embraced after Thomas Aquinas expunged its pagan parts), the heavens were eternal, unchanging, and perfect. Earth sat motionless at the center of God's creation while the sun, moon, stars, and planets revolved around it in perfect circles, carried along by the rotation of the heavenly spheres. According to this cosmology, everything in the terrestrial realm below the moon was corrupted and plagued by rot, death, and decay. The vagaries of life, much like the falling of leaves, were by their very nature fleeting, erratic, and disorderly.

Although an Earth-centered cosmology seemed reassuring and commonsensical, the motion of the planets presented an awkward problem. The word *planet* means "wanderer." In antiquity the planets were known as the wandering stars; instead of maintaining their places in the sky, like the fixed stars in Orion's Belt and the ladle of the Big Dipper, which never moved relative to one another, the planets appeared to drift across the heavens. They progressed from one constellation to another as the weeks and months went by. Most of the time they moved eastward relative to the stars, but occasionally they appeared to slow down, stop, and go backward, westward, in what astronomers called retrograde motion.

Mars, for example, was seen to move in retrograde for about eleven weeks over the course of its nearly two-year circuit around the sky. Nowadays we can capture this reversal photographically. In 2005 the astrophotographer Tunç Tezel took a series of thirty-five snapshots of Mars, each about a week apart, and aligned the images to the stars in the background. In the resulting composite, the eleven dots in the middle show Mars moving in retrograde.

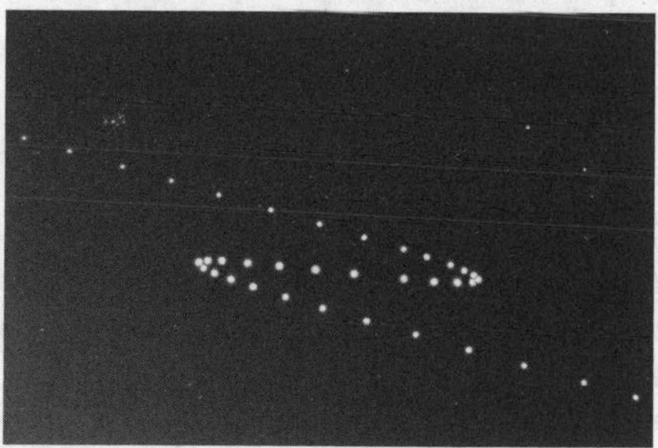

Today we understand that retrograde motion is an illusion. It's caused by our vantage point on Earth as we pass the slower-moving Mars.

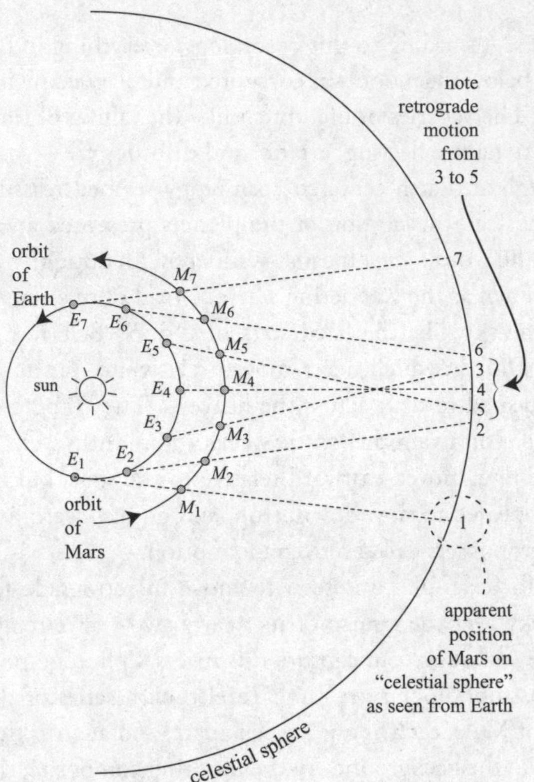

It's like what happens when you pass a car on the highway. Imagine driving on a long highway out in the desert, with mountains off in the distance. As you approach a slower car from behind, it looks like it's moving forward when viewed against the backdrop of the mountains. But when you pull alongside and pass it, the slower car momentarily seems to move *backward* relative to the mountains. Then, once you get far enough ahead of it, the car appears to move forward again.

This kind of observation led the ancient Greek astronomer Aristarchus to propose a sun-centered universe almost two millennia before Copernicus did. It neatly solved the riddle of retrograde motion. However, a sun-centered universe raised questions of its own. If the Earth moves, why don't we fall off? And why do the stars appear fixed? They shouldn't. As the Earth moves around the sun, the distant stars should appear to shift their positions slightly. Experience shows that if you look at something far away and then move and look again, the position of that faraway object appears to shift when viewed against a more distant backdrop. This effect is called parallax. To experience it, hold your finger far out in front of your face. Close one eye, then the other. Your finger seems to shift sideways against the backdrop when you switch eyes. Likewise, as the Earth moves around the sun in its orbit, the stars should shift their apparent positions against the background of even more distant stars. The only way out of this paradox (as Archimedes himself realized when reacting to Aristarchus's sun-centered cosmology) would be if *all* the stars were *immensely* distant, effectively infinitely far away from the Earth. Then the planet's motion would produce no detectable shift, because the parallax would be too small to be measured. This conclusion was hard to accept at the time. No one could imagine a universe so immense with stars so remote, much farther away than the planets. Today we know that is exactly the case, but back then it was inconceivable.

So the Earth-centered cosmology, for all its faults, seemed like the more plausible picture. Suitably modified by the ancient Greek astronomer Ptolemy with epicycles, equants, and other fudge factors, the theory could be made to account reasonably well for planetary motion and it kept the calendar in line with seasonal cycles. The Ptolemaic system was clunky and complicated, but it worked well enough to last into the late Middle Ages.

Two books published in 1543 marked a turning point, the beginning of the scientific revolution. In that year, the Flemish doctor Andreas Vesalius reported the results of his dissections of human cadavers, a practice that had been forbidden in earlier centuries. His findings contradicted fourteen centuries of received wisdom about human anatomy. In that same year, the Polish astronomer Nicolaus Copernicus finally allowed publication of his radical theory that the Earth moved around the sun. He'd waited until he was near death (and died just as the book was being published) because he'd feared that the Catholic Church would be infuriated by his demotion of the world from the center of God's creation. He was right to be scared. After Giordano Bruno proposed, among other heresies, that the universe was infinitely large with infinitely many worlds, he was tried by the Inquisition and burned at the stake in Rome in 1600.

Enter Galileo

Into this climate, as authority and dogma were being challenged by dangerous ideas, Galileo Galilei was born on February 15, 1564, in Pisa, Italy. The eldest son of a once-noble family now down on its luck, Galileo was pushed by his father toward a career in medicine, a much more lucrative profession than his father's own field of music theory. But Galileo soon found his passion was mathematics. He studied Euclid and Archimedes and mastered both. Though he never finished his degree (his family couldn't afford the tuition), he continued to teach himself math and science, got a lucky break as a temporary instructor at Pisa, and gradually rose through the academic ranks as a professor of mathematics at the University of Padua. He was a brilliant lecturer, clear and irreverent with a caustic wit. Students flocked to his classes to hear him.

He met a vivacious and much younger woman named Marina Gamba with whom he had a long and loving but illicit relationship. They had two daughters and a son together but did not marry; it would have been considered dishonorable for him, given Marina's youth and lower social standing. With the strain of his meager salary as a math teacher, the cost of raising their three children, and the additional responsibility to provide for his unwed sister, Galileo felt forced to place his daughters in a convent, which broke his heart. His elder daughter, Virginia, was his favorite, the joy of his life. He later described her as "a woman of exquisite mind, singular goodness, and most tenderly attached to me." When she took her vows as a nun, she chose Sister Maria Celeste as her religious name in honor of the Virgin Mary and her father's fascination with astronomy.

Galileo is perhaps most often remembered today for his work with the telescope and as a champion of the Copernican theory that the Earth moves around the sun, a contradiction of the views of

Aristotle and the Catholic Church. Although Galileo did not invent the telescope, he improved it and was the first to make great scientific discoveries with it. In 1610 and 1611, he observed that the moon had mountains, the sun had spots, and Jupiter had four moons (others have been discovered since then).

All these observations flew in the face of the prevailing dogma. Mountains on the moon meant it was not a glistening, perfect orb, contrary to Aristotelian teaching. Likewise, spots on the sun meant it was not a perfect celestial body; it was marred by blemishes. And since Jupiter and its moons looked like a little planetary system of its own, with four small moons orbiting around a bigger central planet, then clearly not all heavenly bodies revolved solely around the Earth. Furthermore, those moons managed to stay with Jupiter as they all moved across the sky. At the time, one of the standard arguments against heliocentricity was that, if the Earth was orbiting the sun, it would leave the moon behind, but now Jupiter and its moons showed that this reasoning must be false.

This is not to say that Galileo was an atheist or irreligious. He was a good Catholic and believed that he was revealing the glory of God's work by documenting it as it truly was rather than by relying on the received wisdom of Aristotle and his later scholastic interpreters. The Catholic Church, however, did not see it this way. Galileo's writings were condemned as heresy. He was brought before the Inquisition in 1633 and ordered to recant, which he did. He was sentenced to life in prison, a punishment immediately commuted to permanent house arrest in his villa in Arcetri in the hills of Florence. He looked forward to seeing his beloved daughter Maria Celeste, but soon after his return, she fell ill and died, at only thirty-three years of age. Galileo was bereft and for a while lost all interest in work and life.

He spent his remaining years under house arrest, an old man losing his vision and racing against time. Somehow, within two years of his daughter's death, he found the strength within himself to summarize his unpublished investigations of motion from decades earlier. The resulting book, *Discourses and Mathematical Demonstrations Concerning Two New Sciences,* was the culmination of his life's work and the first great masterpiece of modern physics. He wrote it in Italian rather than Latin so that it could be understood by anyone and arranged for it to be smuggled out to Holland, where it was published in 1638. Its radical insights helped launch the scientific revolution and brought humanity to the cusp of discovering the secret of the universe: that the great book of nature is written in calculus.

Falling, Rolling, and the Law of Odd Numbers

Galileo was the first practitioner of the scientific method. Rather than quoting authorities or philosophizing from an armchair, he interrogated nature through meticulous observations, ingenious experiments, and elegant mathematical models. His approach led him to many remarkable discoveries. One of the simplest and most surprising is this: The odd numbers 1, 3, 5, 7, and so forth are hiding in how things fall.

Before Galileo, Aristotle had proposed that heavy objects fall because they are seeking their natural place at the center of the cosmos. Galileo thought these were empty words. Instead of speculating about *why* things fell, he wanted to quantify *how* they fell. To do so, he needed to find a way to measure falling bodies throughout their descent and keep track of where they were moment by moment.

It wasn't easy. Anyone who has dropped a rock off a bridge knows that rocks fall fast. It would take a very accurate clock, of a kind not available in Galileo's day, and several very good video cameras, also not available in the early 1600s, to track a falling rock at each moment of its rapid descent.

Galileo came up with a brilliant solution: He slowed the motion. Instead of dropping a rock off a bridge, he allowed a ball to roll slowly down a ramp. In the jargon of physics, this sort of ramp is known as an inclined plane, although in Galileo's original experiments, it was more like a long, thin piece of wooden molding with a groove cut along its length to act as a channel for the ball. By reducing the slope of the ramp until it was nearly horizontal, he could make the ball's descent as slow as he wished, thus allowing him to measure where the ball was at each moment, even with the instruments available in his day.

To time the ball's descent he used a water clock. It worked like a stopwatch. To start the clock he would open a valve. Water would then flow steadily, at a constant rate, straight down through a thin pipe and into a container. To stop the clock, he would close the valve. By weighing how much water had accumulated during the ball's descent, Galileo could quantify how much time had elapsed to within "one-tenth of a pulse-beat."

He repeated the experiment many times, sometimes varying the tilt of the ramp, other times changing the distances rolled by the ball. What he found, in his own words, was this: "The distances traversed, during equal intervals of time, by a body falling from rest, stand to one another in the same ratio as the odd numbers beginning with unity."

To spell out this law of odd numbers more explicitly, let's suppose the ball rolls a certain distance in the first unit of time. Then, in the next unit of time, it will roll *three* times as far. And in the next unit of time after that, it will roll *five* times as far as it did originally. It's amazing; the odd numbers 1, 3, 5, and so on are somehow inherent in the way things roll downhill. And if falling is just the limit of rolling as the tilt approaches vertical, the same rule must hold for falling.

We can only imagine how pleased Galileo must have been when he discovered this rule. But notice how he phrased it—with words and numbers and proportions, not letters and formulas and equations. Our current preference for algebra over spoken language would have seemed cutting-edge back then, an avant-garde, newfangled way of thinking and speaking. It's not how Galileo would have thought or expressed himself, nor would his readers have understood him if he had.

To see the most important implication of Galileo's rule, let's look at what happens if we add consecutive odd numbers. After one unit of time, the ball has traveled one unit of distance. After the next unit of time the ball has traveled another three units of distance, for a total of 1 + 3 = 4 units traveled since the motion started. After the third unit of time, the total becomes 1 + 3 + 5 = 9 units of distance. Notice the pattern: the numbers 1, 4, and 9 are the squares of consecutive integers — $1^2 = 1$, $2^2 = 4$, $3^2 = 9$. So Galileo's odd-number rule seems to be implying that the total distance fallen is proportional to the square of the time elapsed.

This charming relationship between odd numbers and squares can be proved visually. Think of the odd numbers as L-shaped arrays of dots:

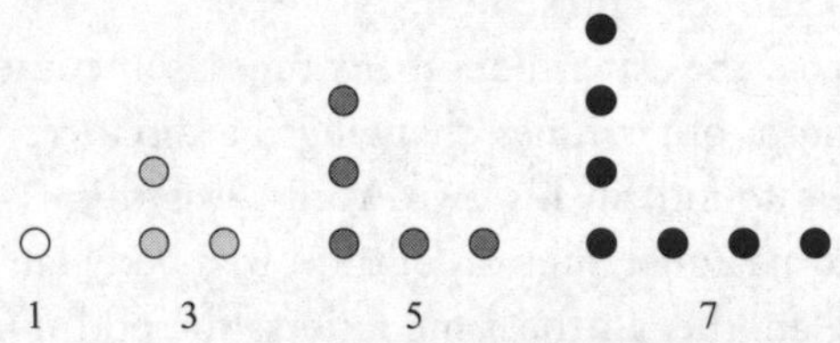

Then nestle them together to form a square. For example, 1 + 3 + 5 + 7 = 16 = 4 × 4, because we can pack the first four odd numbers together to make a 4-by-4 square.

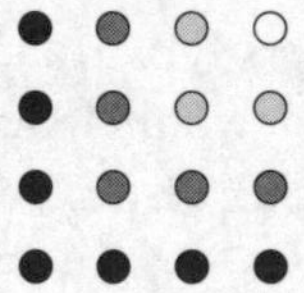

Along with his law about the distance traversed by a falling body, Galileo also discovered a law for its speed. As he put it, the speed increases in proportion to the time of falling. What's interesting about this is that he was referring to the speed of the body at an instant, a seemingly paradoxical concept. He took pains in *Two New Sciences* to explain that when a body falls from rest, it doesn't jump suddenly from zero speed to some higher speed, as his contemporaries thought. Rather, it passes smoothly through every intermediate speed — infinitely many of them — in a finite amount of time, starting from zero and continuously gaining speed as it falls.

So in this law of falling bodies, Galileo was instinctively thinking about *instantaneous speed*, a differential calculus concept that we'll examine in chapter 6. At the time he couldn't make it precise, but he knew what he meant intuitively.

The Art of Scientific Minimalism

Before we leave Galileo's inclined-plane experiment, let's be sure to notice the artistry behind it. He coaxed a beautiful answer out of nature by asking a beautiful question. Like an abstract expressionist painter, he highlighted what he was interested in and cast the rest aside.

For example, in describing his apparatus, he says he made the "groove very straight, smooth, and polished" and "rolled along it a hard, smooth, and very round bronze ball." Why was he so concerned with smoothness, straightness, hardness, and roundness? Because he wanted the ball to roll downhill under the simplest, most ideal conditions he could contrive. He did everything he could to reduce the potential complications coming from friction or from the ball's collisions with the sidewalls of the groove (which could occur if the channel was not straight) or from the ball's softness (which could cause the ball to lose energy if it deformed too much) or from anything else that could cause deviations from the ideal case. Those were the right aesthetic choices. Simple. Elegant. And minimal.

Compare Aristotle, who got the law of falling bodies wrong because he was led astray by complications. He claimed that heavy bodies fell faster than light ones with speeds proportional to their weight. That's true of tiny particles sinking in a very thick, viscous medium like molasses or honey, but not of cannonballs or musket balls dropped through the air. Aristotle seems to have been so concerned with the drag forces produced by air resistance (admittedly an important effect for falling feathers, leaves, snowflakes, and other light objects that also offer an unusual amount of surface area for the air to push up against) that he forgot to test his theory on more typical objects like rocks and bricks and shoes, things that are compact and heavy. In other words, he focused too much on the noise (air resistance) and not enough on the signal (inertia and gravity).

Galileo didn't let himself be distracted. He knew that air resistance and friction were inescapable in the real world, as in his experiment, but they were not of the essence. Anticipating the criticism that he overlooked them in his analysis, he conceded that a pellet of birdshot does not fall quite as fast as a cannonball but noted that the error incurred is much, much less than that produced by Aristotle's theory. In the dialogue of *Two New Sciences*, Galileo's surrogate urges his simple-minded Aristotelian questioner not to "divert the discussion from its main intent and fasten upon some statement of mine which lacks a hair's-breadth of the truth and, under this hair, hide the fault of another which is as big as a ship's cable."

That's the point. In science, being off by a hairsbreadth is acceptable. Being off by a ship's cable is not.

Galileo went on to study projectile motion, like the flight of a musket ball or a cannonball. What sort of arc do they follow? Galileo had the idea that a projectile's motion was compounded of two different effects that could be treated separately: a motion sideways, parallel to the ground, for which gravity played no part, and a vertical motion upward or downward, on which gravity acted and his law of falling bodies applied. Putting those two kinds of motion together, he discovered that projectiles follow parabolic paths. You see them whenever you play a game of catch or take a drink from a water fountain.

This was another stunning connection between nature and math and a further clue that the book of nature is written in the language of mathematics. Galileo was elated to discover that a parabola, an abstract curve studied by his hero Archimedes, was out there in the real world. Nature was using geometry.

To arrive at this insight, however, Galileo again had to know what to neglect. As before, he had to ignore air resistance—the effect of drag on the projectile as it moves through the air. That frictional effect would slow the projectile down. For some kinds of projectiles (a thrown rock), friction is negligible compared to gravity; for others (a beach ball or a Ping-Pong ball), it is not. All forms of friction, including drag caused by air resistance, are subtle and difficult to study. To this day, friction remains mysterious and is a topic of active research.

To get the simple parabola, Galileo needed to assume the sideways motion would continue forever and never slow down. This was an instance of his law of inertia, which states that a body in motion stays in motion at the same speed and in the same direction unless acted on by an outside force. For a real projectile, air resistance would be that outside force. But in Galileo's mind, it was better to start by ignoring it, to capture the lion's share of the truth—and the beauty—of how things move.

From a Swinging Chandelier to the Global Positioning System

Legend has it that Galileo made his first scientific discovery when he was a teenage medical student. One day, while attending a Mass at the Cathedral of Pisa, he noticed a chandelier swaying overhead, moving to and fro like a pendulum. Air currents kept jostling it, and Galileo observed that it always took the same time to complete its swing whether it traversed a wide arc or a small one. That surprised him. How could a big swing and a little swing take the same amount of time? But the more he thought about it, the more it made sense. When the chandelier made a big swing, it traveled farther but it also moved faster. Maybe the two effects balanced out. To test this idea, Galileo timed the swinging chandelier with his pulse. Sure enough, every swing lasted the same number of heartbeats.

This legend is wonderful, and I want to believe it, but many historians doubt it happened. It comes down to us from Galileo's first and most devoted biographer, Vincenzo Viviani. As a young man, he had been Galileo's assistant and disciple near the end of the older man's life, when Galileo was completely blind and under house arrest. In his understandable reverence for his old master, Viviani was known to have embellished a tale or two when he wrote Galileo's biography years after his death.

But even if the story is apocryphal (and it may not be!), we do know for sure that Galileo performed careful experiments with pendulums as early as 1602 and that he wrote about them in 1638 in *Two New Sciences*. In that book, which is structured as a Socratic dialogue, one of the characters sounds like he was right there in the cathedral with the dreamy young student: "Thousands of times I have observed vibrations especially in churches where lamps, suspended by long cords, had been inadvertently set into motion." The rest of the dialogue expounds on the claim that a pendulum takes the same amount of time to traverse an arc of any size. So we know that Galileo was thoroughly familiar with the phenomenon described in Viviani's story; whether he actually discovered it as a teenager is anybody's guess.

In any case, Galileo's assertion that a pendulum's swing always takes the same amount of time is not exactly true; bigger swings take a little longer. But if the arc is small enough—less than 20 degrees, say—it's very nearly true. This invariance of tempo for small swings is known today as the pendulum's *isochronism,* from the Greek words for "equal time." It forms the theoretical basis for metronomes and pendulum clocks, from ordinary grandfather clocks to the towering clock used in London's Big Ben. Galileo himself designed the world's first pendulum clock in the last year of his life, but he died before it could be built. The first working pendulum clock appeared fifteen years later, invented by the Dutch mathematician and physicist Christiaan Huygens.

Galileo was particularly intrigued—and frustrated—by a curious fact he discovered about pendulums, the elegant relationship between its length and its period (the time it takes the pendulum to swing once back and forth). As he explained, "If one wishes to make the vibration-time of one pendulum twice that of another, he must make its suspension four times as long." Using the language of proportions, he stated the general rule. "For bodies suspended by threads of different lengths," he wrote, "the lengths are to each other as the squares of the times." Unfortunately, Galileo never managed to derive this rule mathematically. It was an empirical pattern crying out for a theoretical explanation. He worked at it for years but failed to solve it. In retrospect, he couldn't have. Its explanation required a new kind of mathematics beyond any that he or his contemporaries knew. The derivation would have to wait for Isaac Newton and his discovery of the language God talks, the language of differential equations.

Galileo conceded that the study of pendulums "may appear to many exceedingly arid," although it was anything but that, as later work showed. In mathematics, pendulums stimulated the development of calculus through the riddles they posed. In physics and engineering, pendulums became paradigms of oscillation. Like the line in William Blake's poem about seeing the world in a grain of sand, physicists and engineers learned to see the world in a pendulum's swing. The same mathematics applied wherever oscillations occurred. The worrisome movements of a footbridge, the bouncing of a car with mushy shock absorbers, the thumping of a washing machine with an unbalanced load, the fluttering of venetian blinds in a gentle breeze, the rumbling of the earth in the aftershock of an earthquake, the sixty-cycle hum of fluorescent lights—every field of science and technology today has its own version of to-and-fro motion, of rhythmic return. The pendulum is the granddaddy of them all. Its patterns are universal. *Arid* is not the right word for them.

In some cases, the connections between pendulums and other phenomena are so exact that the same equations can be recycled without change. Only the symbols need to be reinterpreted; the syntax stays the same. It's as if nature keeps returning to the same motif again and again, a pendular repetition of a pendular theme. For example, the equations for the swinging of a pendulum carry over without change to those for the spinning of generators that produce alternating current and send it to our homes and offices. In honor of that pedigree, electrical engineers refer to their generator equations as swing equations.

The same equations pop up yet again, Zelig-like, in the quantum oscillations of a high-tech device that's billions of times faster and millions of times smaller than any generator or grandfather clock. In 1962 Brian Josephson, then a twenty-two-year-old graduate student at the University of Cambridge, predicted that at temperatures close to absolute zero, pairs of superconducting electrons could tunnel back and forth through an impenetrable insulating barrier, a nonsensical statement according to classical physics. Yet calculus and quantum mechanics summoned these pendulum-like oscillations into existence—or, to put it less mystically, they revealed the possibility of their occurrence. Two years after Josephson predicted these ghostly oscillations, the conditions needed to conjure them were set up in the laboratory and, indeed, there they were. The resulting device is now called a Josephson junction. Its practical uses are legion. It can detect ultra-faint magnetic fields a hundred billion times weaker than that of the Earth, which helps geophysicists hunt for oil deep underground. Neurosurgeons use arrays of hundreds of Josephson junctions to pinpoint the sites of brain tumors and locate the seizure-causing lesions in patients with epilepsy. The procedures are entirely noninvasive, unlike exploratory surgery. They work by mapping the subtle variations in magnetic field produced by abnormal electrical pathways in the brain. Josephson junctions could also provide the basis for extremely fast chips in the next generation of computers and might even play a role in quantum computation, which will revolutionize computer science if it ever comes to pass.

Pendulums also gave humanity the first way to keep time accurately. Until pendulum clocks came along, the best clocks were pitiful. They would lose or gain fifteen minutes a day, even under ideal conditions. Pendulum clocks could be made a hundred times more accurate than that. They offered the first real hope of solving the greatest technological challenge of Galileo's era: finding a way to determine longitude at sea. Unlike latitude, which can be ascertained by looking at the sun or the stars, longitude has no counterpart in the physical environment. It is an artificial, arbitrary construct. But the problem of measuring it was real. In the age of exploration, sailors took to the oceans to wage war or conduct trade, but they often lost their way or ran aground because of confusion about where they were. The governments of Portugal, Spain, England, and Holland offered vast rewards to anyone who could solve the longitude problem. It was a challenge of the gravest concern.

When Galileo was trying to devise a pendulum clock in his last year of life, he had the longitude problem firmly in mind. He knew, as scientists had known since the 1500s, that the longitude problem could be solved if one had a very accurate clock. A navigator could set the clock at his port of departure and carry his home time out to sea. To determine the ship's longitude as it traveled east or west, the navigator could consult the clock at the exact moment of local noon, when the sun was highest in the sky. Since the Earth spins through 360 degrees of longitude in a twenty-four-hour day, each hour of discrepancy between local time and home time corresponds to 15 degrees of longitude. In terms of distance, 15 degrees translates to a whopping one thousand miles at the equator. So for this scheme to have any hope of guiding a ship to its desired destination, give or take a few miles of tolerable error, a clock had to run true to within a few seconds a day. And it had to maintain this unwavering accuracy in the face of heaving seas and violent fluctuations in air pressure, temperature, salinity, and humidity, factors that could rust a clock's gears, stretch its springs, or thicken its lubricants, causing it to speed up, slow down, or stop.

Galileo died before he could build his clock and use it to tackle the longitude problem. Christiaan Huygens presented his pendulum clocks to the Royal Society of London as a possible solution, but they were judged unsatisfactory because they were too sensitive to disturbances in their environment. Huygens later invented a marine chronometer whose ticktock oscillations were regulated by a balance wheel and a spiral spring instead of a pendulum, an innovative design that paved the way for pocket watches and modern wristwatches. In the end, however, the longitude problem was solved by a new kind of clock, developed in the mid-1700s by John Harrison, an Englishman with no formal education. When tested at sea in the 1760s, his H4 chronometer tracked longitude to an accuracy of ten miles, sufficient to win the British Parliament's prize of twenty thousand pounds (equivalent to a few million dollars today).

In our own era, the challenge of navigating on Earth still relies on the precise measurement of time. Consider the global positioning system. Just as mechanical clocks were the key to the longitude problem, atomic clocks are the key to pinpointing the location of anything on Earth to within a few meters. An atomic clock is a modern-day version of Galileo's pendulum clock. Like its forebear, it keeps time by counting oscillations, but instead of tracking the movements of a pendulum bob swinging back and forth, an atomic clock counts the oscillations of cesium atoms as they switch back and forth between two of their energy states, something they do 9,192,631,770 times per second. Though the mechanism is different, the principle is the same. Repetitive motion, back and forth, can be used to keep time.

And time, in turn, can determine your location. When you use the GPS in your phone or car, your device receives wireless signals from at least four of the twenty-four satellites in the global positioning system that are orbiting about twelve thousand miles overhead. Each satellite carries four atomic clocks that are synchronized to within a billionth of a second of one another. The various satellites visible to your receiver send it a continuous stream of signals, each of which is time-stamped to the nanosecond. That's where the atomic clocks come in. Their tremendous temporal precision gets converted into the tremendous spatial precision we've come to expect from GPS.

The calculation relies on triangulation, an ancient geolocation technique based on geometry. For GPS, it works like this: When the signals from the four satellites arrive at the receiver, your GPS gadget compares the time they were received to the time they were transmitted. Those four times are all slightly different, because the satellites are at four different distances away from you. Your GPS device multiplies those four tiny time differences by the speed of light to calculate how far away you are from the four satellites overhead. Because the positions of the satellites are known and controlled extremely accurately, your GPS receiver can then triangulate those four distances to determine where it is on the surface of the Earth. It can also figure out its elevation and speed. In essence, GPS converts very precise measurements of time into very precise measurements of distance and thereby into very precise measurements of location and motion.

The global positioning system was developed by the US military during the Cold War. The original intent was to keep track of US submarines carrying nuclear missiles and give them precise estimates of their current locations so that if they needed to launch a nuclear strike, they could target their intercontinental ballistic missiles very accurately. Peacetime applications of GPS nowadays include precision farming, blind landings of airplanes in heavy fog, and enhanced 911 systems that automatically calculate the fastest routes for ambulances and fire trucks.

But GPS is more than a location and guidance system. It allows time synchronization to within a hundred nanoseconds, which is useful for coordinating bank transfers and other financial transactions. It also keeps wireless phone and data networks in sync, allowing them to share the frequencies in the electromagnetic spectrum more efficiently.

I've gone into all this detail because GPS is a prime example of the hidden usefulness of calculus. As is so often the case, calculus operates quietly behind the scenes of our daily lives. In the case of GPS, almost every aspect of the functioning of the system depends on calculus. Think about the wireless communication between satellites and receivers; calculus predicted the electromagnetic waves that make wireless possible through the work of Maxwell that we discussed earlier. Without calculus, there'd be no wireless and no GPS. Likewise, the atomic clocks on the GPS satellites use the quantum mechanical vibrations of cesium atoms; calculus underpins the equations of quantum mechanics and the methods for solving them. Without calculus, there'd be no atomic clocks. I could go on—calculus underlies the mathematical methods for calculating the trajectories of the satellites and controlling their locations and for incorporating Einstein's relativistic corrections to the time measured by atomic clocks as they move at high speeds and in weak gravitational fields—but I hope the main point is clear. Calculus enabled the creation of much of what made the global positioning system possible. Calculus didn't do it on its own, of course. It was a supporting player, but an important one. Along with electrical engineering, quantum physics, aerospace engineering, and all the rest, calculus was an indispensable part of the team.

So let's return to young Galileo sitting in the Cathedral of Pisa pondering that chandelier swinging back and forth. We can see now that his idle thoughts about pendulums and the equal times of their swings had an outsize impact on the course of civilization, not just in his own era but in our own.

Kepler and the Mystery of Planetary Motion

What Galileo did for the motion of objects on Earth, Johannes Kepler did for the motion of the planets in the heavens. He solved the ancient riddle of planetary motion and fulfilled the Pythagorean dream by showing that the solar system was ruled by a kind of celestial harmony. Like Pythagoras with his plucked strings and Galileo with his pendulums, projectiles, and falling bodies, Kepler discovered that planetary motions follow mathematical patterns. And like Galileo, he was enthralled by the patterns he glimpsed and yet frustrated that he couldn't explain them.

Also like Galileo, Kepler was born into a family on the way down. But his circumstances were far worse. His father was a drunken mercenary soldier, "criminally inclined," as Kepler recalled, and his mother was (perhaps understandably) "bad-tempered." On top of that, Kepler contracted smallpox as a child and nearly died from it. His hands and vision were permanently damaged, which meant he could never have a physically strenuous job as an adult.

Fortunately, he was bright. As a teenager he learned mathematics and Copernican astronomy at Tübingen, where he was recognized as having "such a superior and magnificent mind that something special may be expected of him." After receiving his master's degree in 1591, Kepler studied theology at Tübingen and planned to become a Lutheran minister. But when a math teacher at the Lutheran school in Graz died and the church authorities called for a substitute, Kepler was chosen, and he reluctantly gave up the idea of a life in the clergy.

Nowadays, all students of physics and astronomy learn about Kepler's three laws of planetary motion. What is often left out is the story of his agonizing, almost fanatical struggle to uncover those laws. He spent decades toiling, searching for regularities, propelled by mysticism and his faith that there had to be some divine order in the nightly positions of Mercury, Venus, Mars, Jupiter, and Saturn.

A year after his arrival in Graz, a secret of the cosmos was revealed to him, he believed. One day while teaching his class, he suddenly had a vision of how the planets must arrange themselves around the sun. The idea was that the planets were carried by celestial spheres nested inside one another, like Russian dolls, with the distances between them dictated by the five Platonic solids: the cube, tetrahedron, octahedron, icosahedron, and dodecahedron. Plato had known and Euclid had proved that no other three-dimensional shapes could be built from identical regular polygons. To Kepler, their uniqueness and symmetry seemed fit for eternity.

He performed his calculations intensely, feverishly. "Day and night I was consumed by the computing, to see whether this idea would agree with the Copernican orbits, or if my joy would be carried away by the wind. Within a few days everything worked, and I watched as one body after another fit precisely into its place among the planets."

He circumscribed an octahedron about the celestial sphere of Mercury and placed the sphere of Venus through its corners. Then he circumscribed an icosahedron about the sphere of Venus and placed the sphere of Earth through its corners, and so on with the other planets, interlocking the celestial spheres and Platonic solids like a three-dimensional puzzle. He depicted the resulting system in a cutaway drawing in his *Cosmic Mystery* of 1596.

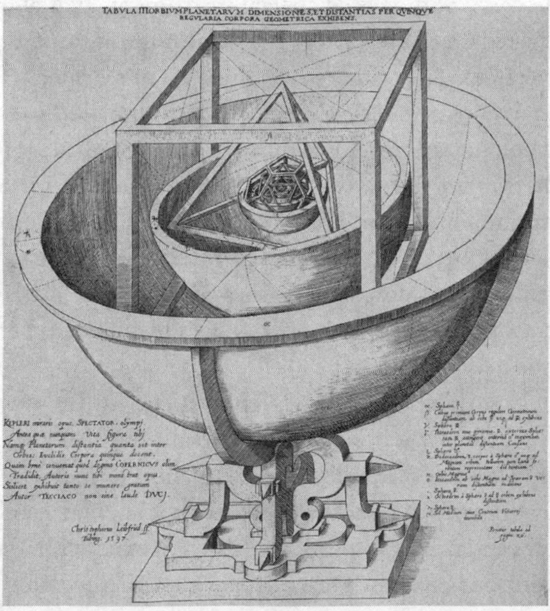

His epiphany explained so much. Just as there were only five Platonic solids, there were only six planets (including the Earth) and hence five gaps between them. Everything made sense. Geometry ruled the cosmos. He had wanted to become a theologian, and now he could write with satisfaction to one of his mentors, "Behold how through my effort God is being celebrated in astronomy."

Actually, the theory didn't quite match the data, particularly as regards the positions of Mercury and Jupiter. That mismatch meant something was wrong, but what was it—his theory, the data, or both? Kepler suspected the data might be wrong, but he didn't insist on the correctness of his theory (which was wise, in retrospect, since the theory had no chance of success; as we now know, there are more than six planets).

Nevertheless, he didn't give up. He continued to ponder the planets and soon got a break when Tycho Brahe asked him to be his assistant. Tycho (as historians always call him) was the world's best observational astronomer. His data were ten times more accurate than any obtained previously. In the days before the invention of the telescope, he'd devised special instruments that allowed him, with the naked eye, to resolve the angular positions of the planets to within two arcminutes. That's one-thirtieth of a degree.

To get a sense of what a tiny angle this is, imagine looking up at the full moon on a clear night while holding your little finger all the way out in front of your face. Your little finger turns out to be about sixty arcminutes wide, and the moon is about half that. So when we say Tycho could resolve two minutes of arc, that means if you drew thirty evenly spaced dots across the width of your little finger (or fifteen across the moon), Tycho could see the difference between one dot and the next.

After Tycho died, in 1601, Kepler inherited his trove of data on Mars and the other planets. To explain their motion, he tried one theory after another, allowing the planets to move in epicycles, in various egg-shaped orbits, and in eccentric circles with the sun slightly off center. But all produced discrepancies with Tycho's data that couldn't be ignored. "Dear reader," he lamented after one such calculation, "if you are tired by this tedious procedure, take pity on me, for I carried it out at least 70 times."

Kepler's First Law: Elliptical Orbits

In his search to explain the motions of the planets, Kepler eventually tried a well-known curve called an ellipse. Like Galileo's parabola, ellipses had been studied in antiquity. As we saw in chapter 2, the ancient Greeks had defined ellipses as the oval-shaped curves formed by cutting through a cone with a plane at a shallow angle, less steep than the slope of the conical surface itself. If the tilt of the cutting plane is shallow, the resulting ellipse is almost circular. At the other extreme, if the tilt of the plane is only slightly less than the tilt of the conical surface, the ellipse is very long and thin, like the shape of a cigar. If you adjust the tilt of the plane, an ellipse can be morphed from very round to very squashed or anywhere in between.

Another way to define an ellipse is in down-to-earth terms and with the help of a few household items.

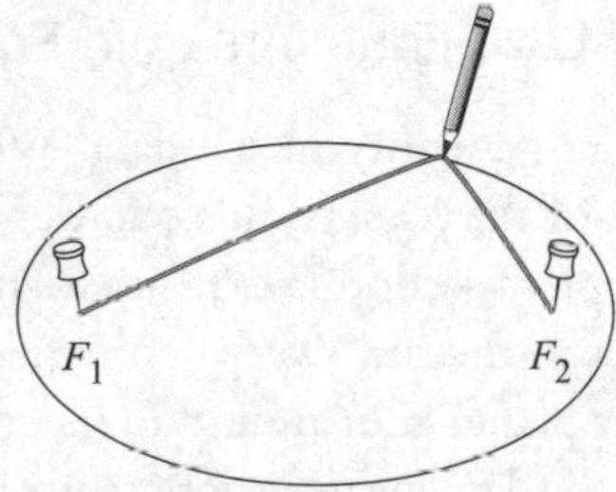

Get a pencil, a corkboard, a sheet of paper, two pushpins, and a piece of string. Place the paper on the corkboard. Pin the ends of the string down through the paper, making sure to leave some slack in the string. Then pull the string taut with the pencil and begin drawing a curve, keeping the string taut as you move the pencil. After the pencil has gone around both pins and returned to its starting point, the resulting closed curve is an ellipse.

The locations of the pins play a special role here. Kepler named them the *foci,* or focal points, of the ellipse. They are as meaningful to an ellipse as the center is to a circle. A circle is defined as a set of points whose distance from a given point (its center) is constant. Likewise, an ellipse is a set of points whose combined distance from *two* given points (its foci) is constant. In the string-and-pushpin construction, that constant combined distance is precisely the length of the loose string between the pins.

Kepler's first great discovery—and this time he really did get it right and didn't need to revise his ideas—is that all the planets move in elliptical orbits. Not circles or circles compounded with circular epicycles, as Aristotle, Ptolemy, Copernicus, and even Galileo had thought. No. Ellipses. Moreover, he found that for every planet, the sun was located at one of the foci of the planet's elliptical orbit.

It was astonishing, just the sort of holy clue Kepler had been hoping for. The planets were moving in accordance with geometry. It hadn't turned out to be the geometry of the five Platonic solids as he'd originally guessed, but his instincts had been right nonetheless. Geometry did rule the heavens.

Kepler's Second Law: Equal Areas in Equal Times

Kepler found another regularity in the data. Whereas the first one was about the paths of the planets, this one was about their speeds. Known today as Kepler's second law, it says that an imaginary line drawn from a planet to the sun sweeps out equal areas in equal intervals of time as the planet goes around in its orbit.

To clarify what this law means, suppose we look at where Mars is tonight in its elliptical orbit. Connect that point to the sun with a straight line.

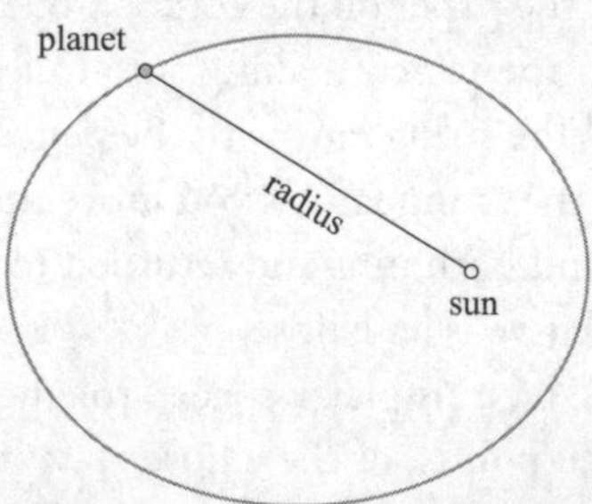

Now think of this line as being something like the blade of a windshield wiper with the sun at the pivot point and Mars at the tip of the wiper (except the wiper doesn't oscillate back and forth like a real windshield wiper; it always advances, and it does so very, very slowly). As Mars travels forward in its orbit on subsequent nights, the wiper moves along with it and thereby sweeps out an area inside the ellipse. If we look at Mars again sometime later, say after three weeks, the slow-moving wiper will have swept out a shape called a sector.

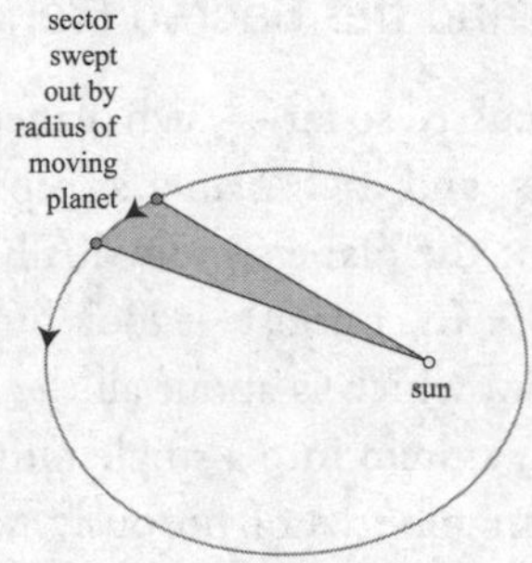

What Kepler discovered is that the area of a three-week sector always stays the same no matter where Mars happens to be in its orbit around the sun. And there's nothing special about three weeks. If we look at Mars at any two points in its orbit separated by equal amounts of time, the resulting sectors will always have equal areas, no matter where they are in the orbit.

In a nutshell, the second law says that the planets do not move at a constant speed. Instead, the closer they get to the sun, the faster they move. The statement about equal areas in equal times is a way of making this precise.

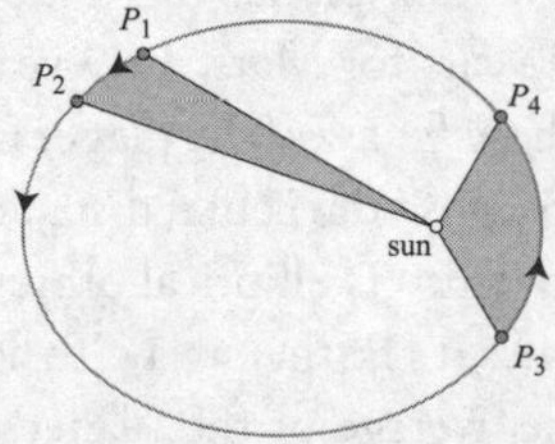

If time ($P_1 \rightarrow P_2$) = time ($P_3 \rightarrow P_4$), their sectors have equal areas.

How did Kepler measure the area of an elliptical sector, given that it had a curved side? He did what Archimedes would have done. He sliced the sector into lots of thin slivers and approximated them with triangles. Next he computed the areas of the triangles (easy, because all their sides are straight) and added them together, integrating them to estimate the area of the original sector. In effect, he used an Archimedean version of integral calculus and applied it to real data.

Kepler's Third Law and the Sacred Frenzy

The laws that we've discussed so far—each planet moves in an ellipse with the sun at a focus, and each planet sweeps out equal areas in equal times—are about the planets individually. Kepler discovered both these laws in 1609. In contrast, it took him another ten years to discover his third law, which is about all the planets collectively. It binds the whole solar system into a single numerological pattern.

It came to him after months of furiously renewed calculations and more than twenty years after his agonizing near miss with the Platonic solids. In the preface to *Harmonies of the World* (1619), he wrote in ecstasy about finally seeing the pattern in God's plan: "Now, since the dawn eight months ago, since the broad daylight three months ago, and since a few days ago, when the full sun illuminated my wonderful speculations, nothing holds me back. I yield freely to the sacred frenzy."

The numerological pattern that enraptured Kepler was his discovery that the square of the period of revolution of a planet is proportional to the cube of its average distance from the sun. Equivalently, the number T^2/a^3 is the same for all the planets. Here, T measures how long it takes a planet to go around the sun once (a year for the Earth, 1.9 years for Mars, 11.9 years for Jupiter, and so on), while a measures how far away the planet is from the sun. That's a bit tricky to define, because the actual distance changes from week to week as a planet moves in its elliptical orbit; sometimes it's closer to the sun and sometimes it's farther away. To account for this effect, Kepler defined a as the average of the planet's nearest and farthest distances to the sun.

The gist of the third law is simple: The farther a planet is from the sun, the slower it moves and the longer it takes to complete its orbit. But what's interesting and subtle about this law is that the orbital period is not simply proportional to the orbital distance. For example, our nearest neighbor, Venus, has a period that's 61.5 percent as long as our year, yet its average distance from the sun is 72.3 percent of ours (not 61.5 percent, as one might naively expect). That's because period *squared* is proportional to distance *cubed* (not squared), and so the relationship between period and distance is more complicated than a direct proportion.

When T and a are expressed as percentages of Earth-years and Earth-distances, as above, Kepler's third law simplifies to $T^2 = a^3$. It becomes an equation instead of a mere proportionality. To see how well it works, plug in the numbers for Venus: $T^2 = (0.615)^2 \approx 0.378$, whereas $a^3 = (0.723)^3 \approx 0.378$. So the law holds to three significant figures. That's what got Kepler so excited. It's equally impressive when applied to the other planets.

Kepler and Galileo, the Same and Not the Same

Kepler and Galileo never met, but they corresponded about their Copernican views and the discoveries they were making in astronomy. When some people refused to look through Galileo's telescope, fearing the instrument was the work of the devil, Galileo wrote to Kepler in a tone of amused resignation: "My dear Kepler, I wish we could laugh at the extraordinary stupidity of the mob. What say you about the foremost philosophers of this University, who with the obstinacy of a stuffed snake, and despite my attempts and invitations a thousand times they have refused to look at the planets, or the moon, or my telescope?"

In some ways, Kepler and Galileo were alike. Both were fascinated by motion. Both worked on integral calculus, Kepler on the volumes of curved shapes, like wine barrels, Galileo on centers of gravity of paraboloids. In this they channeled the spirit of Archimedes, carving solid objects in their minds into imaginary thin wafers, like so many slices of salami.

Yet in other ways, they were complementary to each other. Most obviously, they were complementary in their greatest scientific contributions, Galileo for the laws of motion on Earth, Kepler for the laws of motion in the solar system. But the complementarity goes deeper, down to scientific style and disposition. Where Galileo was rational, Kepler was mystical.

Galileo was the intellectual descendant of Archimedes, entranced by mechanics. In his first publication, he gave the first plausible account of the "Eureka!" legend by showing how Archimedes could have used a balance and a bathtub to determine that King Hiero's crown was not made of pure gold and to calculate the precise amount of silver that the thieving goldsmith had mixed in. Galileo continued to elaborate on Archimedes's work throughout his career, often by extending his mechanics from equilibrium to motion.

Kepler, however, was more the heir to Pythagoras. Fiercely imaginative and with a numerological cast of mind, he saw patterns everywhere. He gave us the first explanation for why snowflakes form six-cornered shapes. He pondered the most efficient way to pack cannonballs, and guessed (correctly) that the optimal packing arrangement is the same one that nature uses to pack pomegranate seeds and that grocers use to stack oranges. Kepler's obsession with geometry, both sacred and profane, verged on the irrational. But his fervor made him who he was. As the writer Arthur Koestler astutely observed, "Johannes Kepler became enamored with the Pythagorean dream, and on this foundation of fantasy, by methods of reasoning equally unsound, built the solid edifice of modern astronomy. It is one of the most astonishing episodes in the history of thought, and an antidote to the pious belief that the Progress of Science is governed by logic."

Storm Clouds Gathering

Like all great discoveries, Kepler's laws of planetary motion in the heavens and Galileo's laws of falling bodies on Earth raised many more questions than they answered. On the scientific side, it was natural to ask about ultimate causes. Where did the laws come from? Did a deeper truth underlie them? For example, it seemed too coincidental that the sun occupied such a special position in all the planetary ellipses, always residing at a focus. Did that mean the sun was affecting the planets somehow? Influencing them through some kind of occult force? Kepler thought so. He wondered if magnetic emanations, recently studied by William Gilbert in England, might be pulling on the planets. Whatever it was, an unknown, invisible force seemed to be acting at great distances across the emptiness of space.

The work of Galileo and Kepler also raised questions for mathematics. In particular, curves were back in the limelight. Galileo had shown that the arc of a projectile was a parabola, and Aristotle's circles had now given way to Kepler's ellipses. Other scientific and technological advances of the early 1600s only heightened the interest in curves. In optics, the shape of a curved lens determined how much an image was magnified, or distorted, or blurred. Those were vital considerations for the design of telescopes and microscopes, the hot new instruments that were revolutionizing astronomy and biology, respectively. The French polymath René Descartes asked: Could a lens be designed to be free of all blurring? It amounted to a question about curves: What curved shape would a lens need to have so that all the rays of light emanating from a single point or traveling parallel to one another would be guaranteed to converge at another unique point after passing through the lens?

Curves, in turn, raised questions about motion. Kepler's second law implied that the planets moved nonuniformly around their ellipses, sometimes hesitating, sometimes accelerating. Likewise, Galileo's projectiles moved at ever-changing speeds on their parabolic arcs. They slowed down as they climbed, paused at the top, then sped up as they fell back to earth. The same was true for pendulums. They slowed down as they climbed to the ends of their arcs, reversed and sped up as they swung through the bottom, then slowed down once again at the other extreme. How could one quantify motions in which speed changed from moment to moment?

Amid this swirl of questions, an influx of ideas from Islamic and Indian mathematics offered European mathematicians a new way forward, a chance to go beyond Archimedes and break new ground. The ideas from the East would lead to fresh ways of thinking about motion and curves and then, with a thunderclap, to differential calculus.

4 The Dawn of Differential Calculus

FROM A MODERN perspective, there are two sides to calculus. Differential calculus cuts complicated problems into infinitely many simpler pieces. Integral calculus puts the pieces back together again to solve the original problem.

Given that cutting comes naturally before rebuilding, it seems sensible for a novice to learn differential calculus first. And indeed, that's how all calculus courses begin today. They start with derivatives—the relatively easy techniques for slicing and dicing—and then work their way up to integrals, the much harder techniques for reassembling the pieces into an integrated whole. Students find it more comfortable to learn calculus in this order because the easier material comes first. Their teachers like it because the subject seems more logical this way.

Yet, strangely enough, history unfolded in the opposite order. Integrals were already in full swing in ancient Greece in Archimedes's work around 250 BCE, whereas derivatives weren't even a gleam in anybody's eye until the 1600s. Why did differential calculus—the easier side of the subject—develop so much later than integral calculus? It's because differential calculus grew out of *algebra*, and algebra took centuries to mature, migrate, and mutate. In its original form in China, India, and the Islamic world, algebra was entirely verbal. Unknowns were words, not today's *x* and *y*. Equations were sentences, and problems were paragraphs. But soon after algebra arrived in Europe, around 1200, it evolved into an art of symbols. That made algebra more abstract . . . and more powerful. This new breed, symbolic algebra, then coupled with geometry and spawned an even stronger hybrid, analytic geometry, which in turn begat a zoo of new curves, the study of which led the way to differential calculus. This chapter explores how that happened.

The Rise of Algebra in the East

The mention of China, India, and the Islamic world should correct the impression I may have given so far that the creation of calculus was a Eurocentric affair. Although calculus culminated in Europe, its roots lie elsewhere. In particular, algebra came from Asia and the Middle East. Its name derives from the Arabic word *al-jabr*, meaning "restoration" or "the reunion of broken parts." These are the kinds of operations needed to balance equations and solve them, such as canceling a number being subtracted from one side of an equation by adding it to both sides, in effect restoring what was broken. Likewise, geometry, as we've seen, was born in ancient Egypt; the founding father of Greek geometry, Thales, is said to have learned the subject there. And the greatest theorem of geometry, the Pythagorean theorem, did not originate with Pythagoras; it was known to the Babylonians for at least a thousand years before him, as evidenced by examples of it on Mesopotamian clay tablets from around 1800 BCE.

We should also keep in mind that when we speak of ancient Greece, we are referring to a huge swath of territory that reached far beyond Athens and Sparta. At its largest, it stretched to Egypt in the south, to Italy and Sicily in the west, and east across the shores of the Mediterranean to Turkey, the Middle East, Central Asia, and parts of Pakistan and India. Pythagoras himself was from Samos, an island off the west coast of Asia Minor (now Turkey). Archimedes lived in Syracuse, on the southeastern coast of Sicily. Euclid worked in Alexandria, the great port and scholarly hub at the mouth of the Nile in Egypt.

After the Romans conquered the Greeks, and especially after the library in Alexandria was burned and the western Roman Empire fell, the center of mathematics swung back to the East. The writings of Archimedes and Euclid were translated into Arabic, as were those of Ptolemy, Aristotle, and Plato. Scholars and scribes in Constantinople and Baghdad kept the old learning alive and added ideas of their own.

How Algebra Waxed While Geometry Waned

During those centuries before algebra arrived, geometry slowed to a crawl. After Archimedes died, in 212 BCE, it seemed that nobody could beat him at his own game. Well, almost nobody. Around 250 CE, the Chinese geometer Liu Hui improved on Archimedes's method for calculating pi. Two centuries later, Zu Chongzhi applied Liu Hui's method to a polygon with 24,576 sides. Through what must have been heroic feats of arithmetic, he tightened the vise on pi to eight digits:

$$3.1415926 < \pi < 3.1415927.$$

The next step forward took another five centuries and came from the sage Al-Hasan Ibn al-Haytham, known to Europeans as Alhazen. Born in Basra, Iraq, around 965 CE, he worked in Cairo during the Islamic golden age on everything from theology and philosophy to astronomy and medicine. In his work on geometry, Ibn al-Haytham calculated volumes of solids that Archimedes never considered. Still, impressive as these advances were, they were rare signs of life for geometry, and they took twelve centuries to occur.

During that same long span of time, rapid and substantial advances were being made in algebra and arithmetic. Hindu mathematicians invented the concepts of zero and the decimal place-value system for numbers. Algebraic techniques for solving equations sprang up in Egypt, Iraq, Persia, and China. Much of this was driven by practical problems involving inheritance law, tax assessment, commerce, bookkeeping, interest calculations, and other topics well suited to numbers and equations. In those days, when algebra was still all about word problems, solutions were given as recipes, step-by-step routes to answers, as elucidated in the famous textbook by Muhammad Ibn Musa al-Khwarizmi (c. 780–850 CE), whose last name lives on in the step-by-step procedures called algorithms. Eventually traders, merchants, and explorers brought this verbal form of algebra and Hindu-Arabic decimals westward to Europe. Meanwhile, people started translating Arabic texts into Latin.

The study of algebra in its own right, as a symbolic system apart from its applications, began to flourish in Renaissance Europe. It reached its pinnacle in the 1500s, when it started to look like what we know today, with letters used to represent numbers. In France in 1591, François Viète designated unknown quantities with vowels, like *A* and *E,* and used consonants, like *B* and *G,* for constants. (Today's use of *x, y, z* for unknowns and *a, b, c* for constants came from the work of René Descartes about fifty years later.) Replacing words with letters and symbols made it much easier to manipulate equations and find solutions.

An equally big advance in the realm of arithmetic came when Simon Stevin in Holland showed how to generalize Hindu-Arabic decimal numbers to decimal fractions. In so doing, he destroyed the old Aristotelian distinction between numbers (meaning whole numbers of indivisible units) and magnitudes (continuous quantities that could be divided infinitely into arbitrarily small parts). Before Stevin, decimals had been applied only to the whole-number part of a quantity, and any part less than a unit was expressed as a fraction. In Stevin's new approach, even a unit could be chopped into pieces and written in decimal notation by placing the correct digits after the decimal point. It sounds simple to us now, but it was a revolutionary idea that helped make calculus possible. Once the unit was no longer sacrosanct and indivisible, all quantities—whole, fractional, or irrational—coalesced into one big family of numbers, all on equal footing. That gave calculus the infinitely precise real numbers it needed to describe the continuity of space, time, motion, and change.

Just before geometry partnered with algebra, there was one last hurrah for the old-school geometric methods of Archimedes. At the beginning of the seventeenth century, Kepler found the volumes of curved shapes like wine barrels and doughnut-shaped solids by slicing them in his mind into an infinite number of infinitesimally thin disks, while Galileo and his students Evangelista Torricelli and Bonaventura Cavalieri similarly computed areas, volumes, and centers of gravity of various shapes by treating them as infinite stacks of lines and surfaces. Because these men had a devil-may-care approach to infinity and infinitesimals, their techniques were not rigorous, but they were potent and intuitive. They produced answers much more easily and quickly than the method of exhaustion, so this seemed like an exciting advance (though we now know that Archimedes had beaten them to it; the same idea lay hidden in his treatise on the Method, at that time still languishing undetected in a prayer book in a monastery, where it would remain until 1899).

At any rate, although the progress made by the neo-Archimedeans seemed promising at the time, this continuation of the old approach was not destined to carry the day. Symbolic algebra was now where the action was. And with it, the seeds for its most vigorous offshoots—analytic geometry and differential calculus—were finally about to be sown.

Algebra Meets Geometry

The first breakthrough came around 1630 when two French mathematicians (and soon-to-be rivals), Pierre de Fermat and René Descartes, independently linked algebra to geometry. Their work created a new kind of mathematics, analytic geometry, whose central theater was the *xy* plane, an arena where equations came alive and took form.

We use the *xy* plane today to graph relationships between variables. For example, consider the caloric implications of my occasionally disgraceful eating habits. Sometimes I treat myself to a couple of slices of cinnamon-raisin bread for breakfast. On the package it tells me that each slice packs a whopping 200 calories. (If I wanted to eat healthier, I could always settle for the seven-grain bread my wife buys, with its 130 calories per slice, but for this example, I prefer the cinnamon-raisin bread because 200 is a more congenial number, mathematically if not nutritionally, than 130.)

Here's a graph of how many calories I consume when I eat one, two, or three slices of bread.

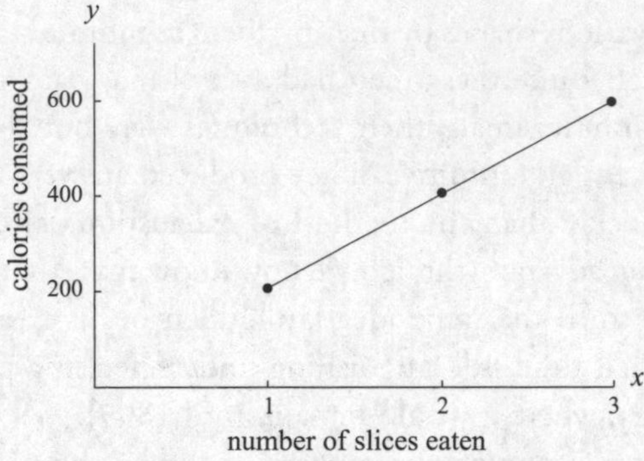

Since each slice amounts to 200 calories, two slices amount to 400 calories, and three to 600 calories. When plotted as data points on the graph, all three points fall on the same straight line. In that sense, there's a *linear* relationship between calories consumed and number of slices eaten. If we use the letter x to represent the number of slices eaten and y for the number of sinful calories ingested, the linear relationship can be summarized as $y = 200x$. This relationship also applies between the data points. For example, one and a half slices amount to 300 calories, and the corresponding data point falls right on the line. So it makes sense to connect the dots in graphs like these.

I realize all of this might seem obvious, but that's my point. It wasn't always obvious. It wasn't obvious in the past—someone had to come up with the idea to depict relationships on an abstract visual chart—and it still isn't obvious today, at least not to kids when they first learn about graphs like this.

There are several imaginative leaps here. One is to use a picture to represent food intake. That requires mental flexibility. There is nothing inherently pictorial about calories. The graph we're looking at is not a photorealistic painting showing raisins and brown swirls of cinnamon embedded in bread. The graph is an abstraction. It allows different mathematical domains to interact and cooperate: the domain of numbers, like numbers of calories or slices of bread; the domain of symbolic relationships, like $y = 200x$; and the domain of shapes, like dots lying on a straight line on a graph with two perpendicular axes. Through this confluence of ideas, the humble chart blends numbers, relationships, and shapes and hence lets arithmetic and algebra merge into geometry. That's the big deal here. Different streams of mathematics have been brought together after centuries of running on their separate courses. (Recall that the ancient Greeks elevated geometry over arithmetic and algebra and didn't let them mingle, at least not very often.)

Another confluence here involves the horizontal and vertical axes. They are often called the x and y axes, named for the variables we use to label them. These axes are number lines. Think about that term: *number lines. Numbers* are being represented as *points* on a line. Arithmetic is consorting with geometry. And they're mingling before we even plot any data!

The ancient Greeks would have screamed bloody murder at that breach of protocol. To them, numbers meant exclusively discrete quantities, like whole numbers and fractions. By contrast, continuous quantities of the sort measured by the length of a line were regarded as magnitudes, a conceptually distinct category from numbers. So for the nearly two thousand years from Archimedes to the beginning of the seventeenth century, numbers were absolutely *not* seen as equivalent to the continuum of points on a line. In this sense, the idea of a number line was radically transgressive. Nowadays we don't give it a second thought. We expect elementary-school children to understand that numbers can be represented visually in this way.

Further blasphemy here, from the standpoint of the ancient Greeks, is the graph's utter disregard for comparing like with like, apples with apples or calories with calories. Instead, the graph shows calories on one axis and slices on the other. They are not directly comparable. And yet we don't blink an eye at making such comparisons today when we draw graphs like this. We simply convert calories and slices to numbers, meaning real numbers, infinite decimals, the universal currency of continuous mathematics. The Greeks drew sharp distinctions between lengths, areas, and volumes, but they're all just real numbers to us.

Equations as Curves

To be sure, Fermat and Descartes never used the *xy* plane to study anything as tangible as cinnamon-raisin bread. For them, the *xy* plane was a tool to study pure geometry.

Working separately, they each noticed that any linear equation (meaning an equation in which x and y appear to the first power only) produced a straight line on the *xy* plane. This connection between linear equations and lines suggested the possibility of a deeper connection, one between *nonlinear* equations and *curves*. In a linear equation like $y = 200x$, the variables x and y appear on their own, unadulterated, and do not get squared or cubed or raised to any higher power. Fermat and Descartes realized they could play the same game with other powers and other equations. They could cook up any equation they desired and do whatever they wanted to x and y—square one of them, cube the other, multiply them together, add them, whatever—and then interpret the result as a curve. With any luck, it would be an interesting curve, maybe one that nobody had ever imagined, maybe one that Archimedes had never studied. Any equation with x and y in it was a new adventure. It was also a gestalt switch. Instead of starting with a curve, you start with an equation and see what kind of curve it makes. Let algebra drive, and put geometry in the back seat.

Fermat and Descartes began by looking at quadratic equations. These are equations in which, along with the usual constants (like 200) and linear terms (like x and y), the variables can also get squared or multiplied together, creating quadratic terms like x^2, y^2, and xy. (In Latin, *quadratus* means "square.") Squared quantities had traditionally been interpreted as the areas of square regions. Thus, x^2 meant the area of an x-by-x square. In the old days, an area was seen as a fundamentally different kind of quantity from a length or a volume. But to Fermat and Descartes, x^2 was just another real number, which meant it could be graphed on a number line, just as x or x^3 or any other power of x could be.

Today, students in high-school algebra are routinely expected to be able to graph equations like $y = x^2$, whose associated curve turns out to be a parabola. Remarkably, all other equations involving quadratic terms in x and y but no higher powers give curves of just four possible types: parabolas, ellipses, hyperbolas, or circles. And that's it. (Except for some degenerate cases that yield lines, points, or no graph at all, but these are rare oddities that we can safely ignore.) For example, the quadratic equation $xy = 1$ gives a hyperbola, while $x^2 + y^2 = 4$ is a circle and $x^2 + 2y^2 = 4$ is an ellipse. Even a quadratic as beastly as $x^2 + 2xy + y^2 + x + 3y = 2$ has to be one of the four possibilities above. It turns out to be a parabola.

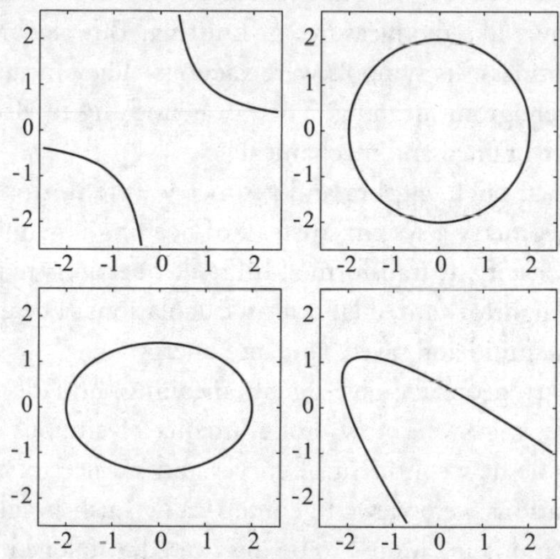

Fermat and Descartes were the first to discover this wonderful coincidence: The quadratic equations in x and y are the algebraic counterparts of the conic sections of the Greeks, the four kinds of curves obtained by slicing through a cone at different angles. Here, in Fermat's and Descartes's new arena, classical curves were reappearing like ghosts from the mist.

Better Together

The newfound ties between algebra and geometry were a boon to both subjects. Each could help the other compensate for its deficits. Geometry appealed to the right side of the brain. It was intuitive and visual, and the truths of its propositions were often clear at a glance. But it called for a certain kind of ingenuity. With geometry, there was often no clue about where to start a proof. Beginning an argument required strokes of genius.

Algebra, however, was systematic. Equations could be massaged almost mindlessly, peacefully; you could add the same term to both sides of an equation, cancel common terms, solve for an unknown quantity, or perform a dozen other procedures and algorithms according to standard recipes. The processes of algebra could be soothingly repetitive, like the pleasures of knitting. But algebra suffered from its emptiness. Its symbols were vacuous. They meant nothing until they were given meaning. There was nothing to visualize. Algebra was left-brained and mechanical.

Together, though, algebra and geometry were unstoppable. Algebra gave geometry a system. Instead of needing ingenuity, it now demanded tenacity. It transformed difficult questions requiring insight into straightforward, if laborious, calculations. The use of symbols freed the mind and saved time and energy.

For its part, geometry gave algebra meaning. Equations were no longer sterile; they were now embodiments of sinuous geometric forms. A whole new continent of curves and surfaces opened up as soon as equations were viewed geometrically. Lush jungles of geometric flora and fauna waited to be discovered, cataloged, classified, and dissected.

Fermat Versus Descartes

Anyone who has studied a lot of math and physics will have run into the names of Fermat and Descartes. But none of my teachers or textbooks ever told me about their rivalry or how vicious Descartes could be. To understand what was at stake in their fights, you need to know more about their lives, their personalities, and what they hoped to achieve.

René Descartes (1596–1650) was one of the most ambitious thinkers of all time. Daring, intellectually fearless, and contemptuous of authority, he had an ego as big as his genius. For example, of the Greek approach to geometry, which all other mathematicians had revered for two thousand years, he wrote dismissively: "What the ancients have taught us is so scanty and for the most part so lacking in credibility that I may not hope for any kind of approach toward truth except by rejecting all the paths which they have followed." At a personal level, he could be paranoid and thin-skinned. The most famous portrait of him shows a man with a gaunt face, haughty eyes, and a snide little mustache. He looks like a cartoon villain.

Descartes set out to rebuild human knowledge on a foundation of reason, science, and skepticism. He is best known for his work in philosophy, immortalized by his famous line *Cogito, ergo sum* ("I think, therefore I am"). In other words, when all is in doubt, at least one thing is certain: the doubting mind exists. His analytic approach, which appears to have been inspired by the rigorous logic of mathematics, is widely seen today as the beginning of modern philosophy. In his most famous book, his *Discourse on Method*, Descartes introduced a bracing new style of thinking about philosophical problems, but he also included three appendices of interest in their own right—one on geometry, in which he presented his approach to analytic geometry; another on optics, of great import at a time when telescopes, microscopes, and lenses were the latest technology; and a third on weather, which has mostly been forgotten except for his correct explanation of rainbows. His capacious intellect roamed far and wide. He viewed the living body as a system of mechanical devices and located the seat of the soul in the brain's pineal gland. He proposed a grand (but wrong) system of the universe according to which invisible vortices pervaded all space, with the planets carried along like leaves in a whirlpool.

Born into a wealthy family, Descartes was sickly as a little boy and was allowed to stay in bed and read and think as long as he liked, a habit he kept his whole life, never rising before noon. His mother died when he was just a year old, but fortunately she left him a sizable inheritance that later allowed him to live a life of leisure and adventure as a wandering gentleman. He volunteered for the Dutch army but never saw combat, and he had plenty of time for philosophy. He spent much of his adult life in Holland, working on his ideas and corresponding and bickering with other great thinkers. In 1650, he reluctantly took a position in Sweden (which he scorned as "the country of bears, amid rocks and ice") as Queen Christina's personal philosophy tutor. Unfortunately for Descartes, the energetic young queen was an early riser. She insisted on lessons at five in the morning, an ungodly hour for anyone but especially for Descartes, accustomed to getting up at noon his whole life. That winter in Stockholm was the coldest in decades. After a few weeks, Descartes caught pneumonia and died.

Pierre de Fermat (1601–1665), who was five years younger than Descartes, lived a peaceful, upper-middle-class, comparatively uneventful life. By day he was a lawyer and provincial judge in Toulouse, far from the hubbub in Paris. By night he was a husband and father. He came home from work, ate dinner with his wife and five kids, and then spent a few hours with his one true passion: doing math. Whereas Descartes was a big thinker of colossal ambitions, Fermat was a shy man, quiet, even-tempered, and naive. He had more modest goals than Descartes did. He didn't see himself as a philosopher or a scientist. Math was enough for him. He pursued it as an amateur, lovingly. He saw no need to publish, and he didn't. He wrote little notes to himself in the books he was reading, classic Greek tomes by Diophantus and Archimedes, and occasionally mailed his ideas to scholars he thought might appreciate them. He never traveled far from Toulouse or met any of the major mathematicians of his day, although he corresponded with them through Marin Mersenne, a Franciscan friar, mathematician, and social connector.

It was through Mersenne that Fermat and Descartes locked horns. Among mathematicians, Mersenne was the go-to guy in Paris. In a time before Facebook, he kept everyone in touch with everyone else, a real busybody with a certain lack of tact and discretion. He had a way of stirring up trouble; for example, he showed people personal letters he received and released confidential manuscripts before they were published. There was a circle around him of top mathematicians, not quite in the same league as Fermat and Descartes, but strong nonetheless, and they apparently had it in for Descartes. They were always sniping at him and his grandiose *Discourse on Method.*

So when Descartes heard via Mersenne that some nobody in Toulouse, some amateur named Fermat, claimed to have developed analytic geometry a decade earlier than he had and that this same amateur (who *was* this guy?) had raised doubts about his theory of optics, Descartes considered it another case of someone out to get him. In the years to come, he fought vehemently against Fermat and tried to ruin his reputation. After all, Descartes had a lot to lose. In the *Discourse,* he'd claimed that his analytical method was the one sure route to knowledge. If Fermat could outdo him without even using his method, his whole project was at risk.

Descartes badmouthed Fermat mercilessly and to some extent succeeded in diminishing him. Fermat's work was never properly published until 1679. His results trickled out through word of mouth or in copies of his letters, but he was not truly appreciated until long after his death. Descartes, however, hit it big. His *Discourse* became famous. The next generation learned analytic geometry from it. Even today, our students learn about Cartesian coordinates, even though Fermat came up with them first.

The Search for Analysis, the Long-Lost Method of Discovery

The squabbles between Descartes and Fermat took place against the backdrop of the early seventeenth century, a time when mathematicians dreamed of finding a method of analysis for geometry. Here *analysis,* as in analytic geometry, is to be understood in the archaic sense of the word—as a means of discovering results rather than proving them. There was widespread suspicion at the time that the ancients had possessed such a method of discovery but had deliberately concealed it. Descartes, for example, alleged that the ancient Greeks "had knowledge of a species of mathematics very different from that which passes current in our time . . . but my opinion is that these writers then, with a sort of low cunning, deplorable indeed, suppressed this knowledge."

Symbolic algebra seemed like it might be this long-lost method of discovery. But in more conservative quarters, algebra met with reactionary skepticism. A generation later, when Isaac Newton said, "Algebra is the analysis of the bunglers in mathematics," he was throwing a thinly veiled insult at Descartes, the prime example of a "bungler" who had relied on algebra as a crutch to solve problems by working backward.

In launching his attack, Newton was adhering to a traditional distinction between analysis and synthesis. In analysis, one solves a problem by starting at the end, as if the answer had already been obtained, and then works back wishfully toward the beginning, hoping to find a path to the given assumptions. It's what kids in school think of as working backward from the answer to figure out how to get there.

Synthesis goes in the other direction. It starts with the givens, and then, by stabbing in the dark, trying things, you are somehow supposed to move forward to a solution, step by logical step, and eventually arrive at the desired result. Synthesis tends to be much harder than analysis because you don't ever know how you're going to get to the solution until you do.

The ancient Greeks regarded synthesis as carrying more logical force, more persuasive power, than analysis. Synthesis was considered the only valid way to *prove* a result; analysis was a practical way to *find* the result. If you wanted a rigorous demonstration, you had to do synthesis. That's why, for example, Archimedes used his analytical method of balancing shapes on seesaws to find his theorems but then switched to the synthetic method of exhaustion to prove them.

Still, although Newton looked down his nose at algebraic analysis, we will see in chapter 7 that he used it himself, and to tremendous effect. But he wasn't its first master. Fermat was. Fermat's style of thinking is fun to examine, because it's elegant and accessible and yet foreign and surprising. His methods for studying curves are no longer in use, having been superseded by the more sophisticated techniques in the textbooks today.

Optimizing for the Overhead Bin

Fermat's embryonic version of differential calculus grew out of his application of algebra to optimization problems. Optimization is the study of how to do things in the best possible way. Depending on context, *best* might mean fastest, cheapest, biggest, most profitable, most efficient, or some other notion of optimality. To illustrate his ideas in the simplest fashion, Fermat contrived a few problems that sound a lot like the exercises we math teachers are still assigning to our students today. They can blame it all on him.

One of those problems, updated for our time, goes something like this. Imagine you want to design a rectangular box to hold as much stuff as possible, subject to two constraints. First, the box has to have a square cross section, x inches wide by x inches deep. Second, it has to fit in the overhead bin of a certain airline. According to their rules for carry-on baggage, the width plus depth plus height of the box cannot exceed 45 inches. What choice of x produces a box of maximum volume?

One way to solve this is with common sense. Try a few possibilities. Say the width and depth are 10 inches each. That would allow for a height of 25 inches, since $10 + 10 + 25 = 45$. A box with those dimensions would have a volume of $10 \times 10 \times 25$, which equals 2,500 cubic inches. Would a cube-shaped box be better? Since a cube must have equal height, width, and depth, it would have to have dimensions $15 \times 15 \times 15$, which multiplies out to a roomy 3,375 cubic inches. Fiddling around with a few other possibilities makes it seem likely that a cube is the optimal choice for the shape of the box. And indeed it is.

So this is not a particularly hard problem in itself. The point of it is to show how Fermat reasoned about such problems, because his approach led to much greater things.

As in most algebra problems, the first step is to translate all the given information into symbols. Since the width and depth of the box are both x, they add up to $2x$. And since the height plus width plus depth cannot exceed 45 inches, that leaves $45 - 2x$ for the height. Thus the volume will be x times x times $(45 - 2x)$. Multiplying that out gives $45x^2 - 2x^3$. That's the volume of our box. Call it $V(x)$. Thus

$$V(x) = 45x^2 - 2x^3.$$

If we cheat momentarily and use a computer or a graphing calculator to plot x horizontally and V vertically, we see that the curve rises up and reaches its maximum when $x = 15$ inches, as expected, and then descends back to zero.

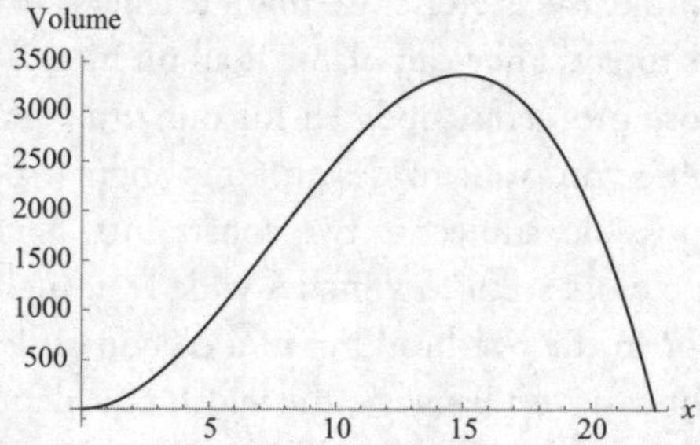

Alternatively, to find that maximum with differential calculus as practiced today, our students would reflexively take the derivative of V and set it equal to zero. The thinking is that at the top of the curve, the slope is zero. The curve is neither rising nor falling there. So, since the slope is measured by the derivative (as we'll see in chapter 6), the derivative must be zero at the maximum. After a bit of algebra and the incantation of various memorized rules for derivatives, this line of reasoning would also yield $x = 15$ at the maximum.

But Fermat didn't have graphical calculators or computers, and he certainly didn't have the concept of derivatives; on the contrary, he invented the ideas that *led* to derivatives! So how did he solve the problem? He used a special property of the maximum: horizontal lines below the maximum intersect the curve at two points, as shown here,

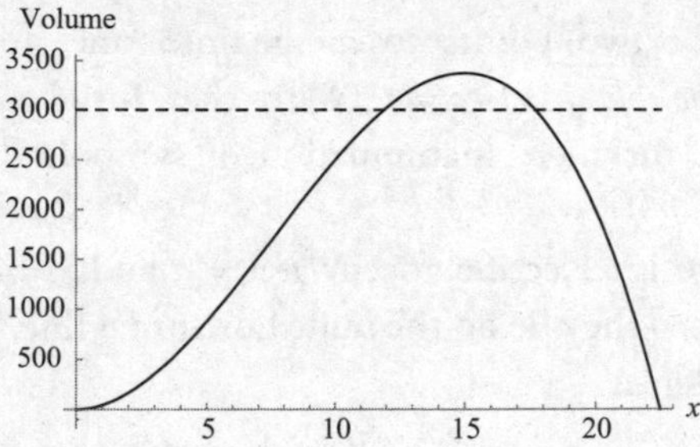

whereas horizontal lines above the maximum don't intersect the curve at all.

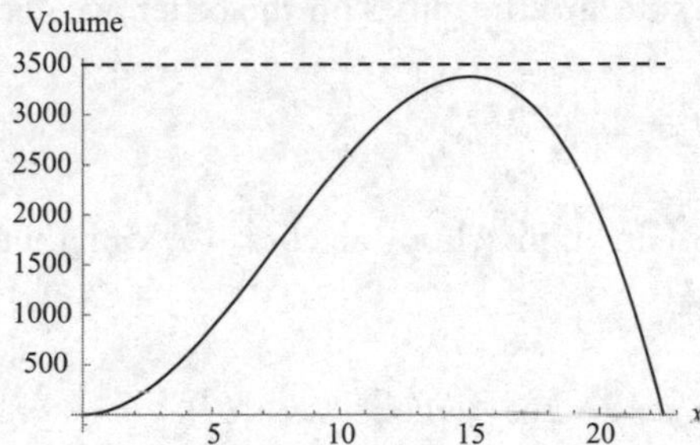

That suggested an intuitive strategy to solve the problem. Imagine slowly lifting a horizontal line that starts below the maximum. As the line gradually moves up, its two intersection points slide toward each other along the curve like beads on a necklace.

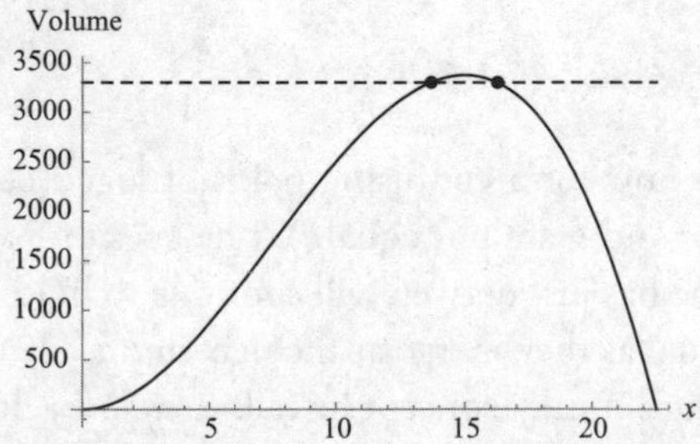

At the maximum, those two points collide. Looking for that collision was how Fermat determined the maximum. He derived a condition for two points to merge into one, forming what's known as a *double intersection.* With that insight in place, the rest is algebra, the mere manipulation of symbols. It goes as follows.

Say the two intersections occur at $x = a$ and $x = b$. Then since (by construction) they lie on the same horizontal line, we must have $V(a) = V(b)$. Hence

$$45a^2 - 2a^3 = 45b^2 - 2b^3.$$

To make headway, it helps to rearrange this equation. If we put the squares on one side and the cubes on the other, we get

$$45a^2 - 45b^2 = 2a^3 - 2b^3.$$

With some skill in high-school algebra, we can then factor both sides to obtain

$$45(a - b)(a + b) = 2(a - b)(a^2 + ab + b^2).$$

Next, divide both sides by the common factor of $a - b$. That's legal, since a and b are assumed to be different. (If they were equal, dividing both sides by $a - b$ would amount to dividing by zero, which is prohibited, as discussed in chapter 1.) After cancellation, the resulting equation is

$$45(a + b) = 2(a^2 + ab + b^2).$$

Buckle up now for a confusing point of logic. Fermat has just assumed that a and b are not equal. Yet he goes on to imagine that the equation he has just derived will continue to hold when a and b do become equal as they merge at the maximum. He tries to justify this by invoking a murky concept he calls *adequality*. It expresses the idea that a and b become sort of equal but not really equal at the maximum (today we would phrase it using the concept of a limit or a double intersection). Anyway, he sets $a \approx b$, where the squiggly equals sign means adequal, and then cavalierly substitutes a for b in the equation above to get

$$45(2a) = 2(a^2 + a^2 + a^2).$$

This simplifies to $90a = 6a^2$, whose solutions are $a = 0$ and $a = 15$. The first of these solutions, $a = 0$, gives a box of *minimum* volume; it has zero width and depth and hence has zero volume. That's of no interest. The second solution, $a = 15$, gives the box of *maximum* volume. There's the answer we've been expecting: 15 inches is the optimal width and depth.

From today's perspective, Fermat's reasoning seems strange. He finds a maximum without using derivatives. Today we teach derivatives before optimization; Fermat did it the other way around. But it doesn't matter. His ideas are equivalent to ours.

How Fermat Helped the FBI

The legacy of Fermat's early work on optimization is all around us. Our lives today depend on algorithms that solve optimization problems using the notion of double intersections and equivalent conditions expressed with derivatives. Today's problems tend to be much more complicated than Fermat's, but the spirit is the same.

One important application involves big data sets, where it's often helpful to code the data as compactly as possible. For example, the Federal Bureau of Investigation has millions of records of fingerprints. To store them, search them, and retrieve them efficiently for background checks, they use calculus-based methods of data compression. Clever algorithms reduce the size of the digitized fingerprint files without sacrificing any details that matter. The same is true when you store music and pictures on your phone. Rather than keep every note and pixel, compression algorithms named MP3 and JPEG save space by distilling the information down to a much more efficient form. They also let us download songs and photos quickly and send them to our loved ones without clogging up their inboxes too much.

To see what calculus and optimization have to do with data compression, let's take a look at the related statistical problem of fitting a curve to data, an issue that comes up everywhere from climate science to business forecasting. The data set we'll examine shows how day length varies with the seasons. As we all know, the days are longer in the summer and shorter in the winter, but what does the overall pattern look like? In the graph below, I've plotted the data for New York City for the year 2018, with time running horizontally from January 1 on the far left to December 31 on the far right. The vertical axis shows the number of minutes between sunrise and sunset at different times of the year. To avoid cluttering up the picture, I've shown the data for only twenty-seven days, sampled every two weeks starting on January 1.

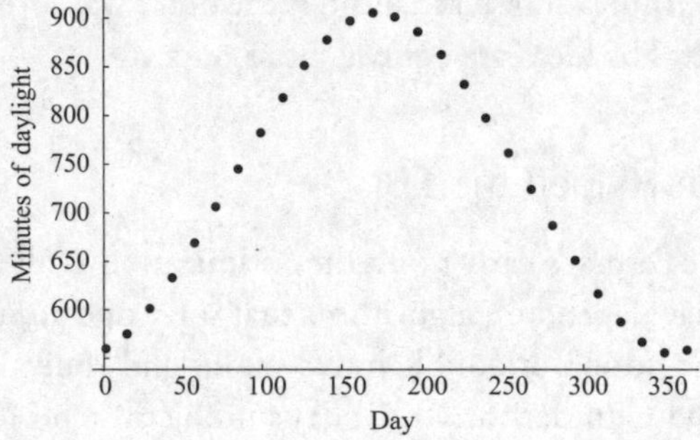

The graph shows that day length rises and falls throughout the year, as expected. The days are longest around the summer solstice (June 21, corresponding to the peak at day 172 near the middle of the graph) and shortest around the winter solstice, half a year later. Overall the data appear to lie on a smoothly undulating wave.

In high-school trigonometry classes, teachers talk about a certain kind of wave, a sine wave. Later in this book I will have more to say about what sine waves are and why they are special from the standpoint of calculus. For now, the main thing we need to know is that sine waves are connected to circular motion. To see the connection, imagine a point moving around a circle at a constant speed. If we track its up-and-down position as a function of time, the point traces out a sine wave.

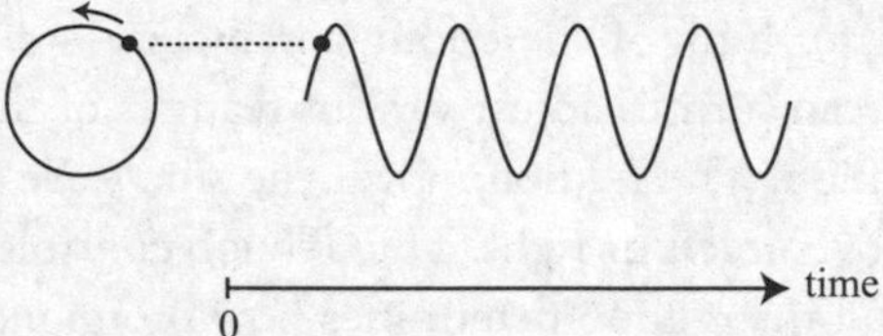

And because circles are intimately connected to cycles, sine waves come up wherever cyclic phenomena occur, from the cycle of the seasons to the vibrations of a tuning fork to the sixty-cycle hum of fluorescent lights and power lines. That annoying hum is the sound of sine waves bobbing up and down sixty times a second. It's the telltale sign of alternating current produced by generators in the power grid whose machinery is spinning at that same frequency. Where there is circular motion, there are sine waves.

Any sine wave is completely defined by four vital statistics: its period, average, amplitude, and phase.

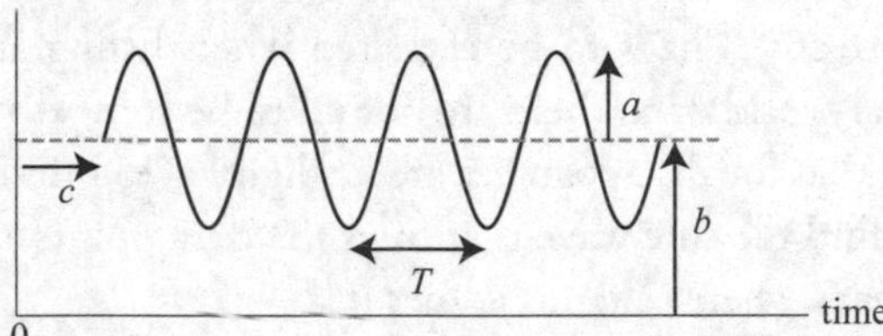

These four parameters have simple interpretations. The period T indicates how long it takes the wave to complete a full cycle. For the day-length data we're considering here, T is about a year or, to be more precise, 365.25 days. (That extra quarter of a day is why we need leap years every fourth year, to keep the calendar in sync with natural cycles.) The average of the sine wave is its baseline value, b. For our data, it's the typical number of minutes of daylight in New York City averaged across all the days of the year 2018. The wave's amplitude a tells us how many additional minutes of light there are on the longest day of the year as compared to the average day. The wave's phase c tells us the day on which the wave crosses upward through its average value, sometime around the spring equinox.

It's helpful to think of these four parameters—a, b, c, T—as four knobs we can turn to adjust various features of the sine wave's shape and location. The b-knob moves the sine wave up or down. The c-knob moves it left or right. The T-knob controls how rapidly it oscillates. And the a-knob determines how pronounced those oscillations are.

If we could somehow set the knobs to make the sine wave go through all the data points we plotted earlier, that would amount to a significant compression of information. It would mean we were capturing the twenty-seven data points with just the four parameters in the sine wave, thereby compressing the data by a factor of $27/4$, or 6.75. Actually, since we know one of the parameters is a year, we really have only three parameters to fiddle with, giving us a compression factor of $27/3$, or 9. A reduction of this size is conceivable because the data are not random. They follow a pattern. The sine wave embodies that pattern and does the work for us.

The only catch is that there is no sine wave that goes through the data perfectly. That's to be expected when fitting an idealized model to real-world data; there are bound to be some discrepancies. The hope is that the discrepancies are negligible. To minimize them, we need to find the sine wave that hugs the data points as closely as possible. That's where calculus comes in.

The figure below shows the best-fitting sine wave, as determined by an optimization algorithm I'll explain in a minute.

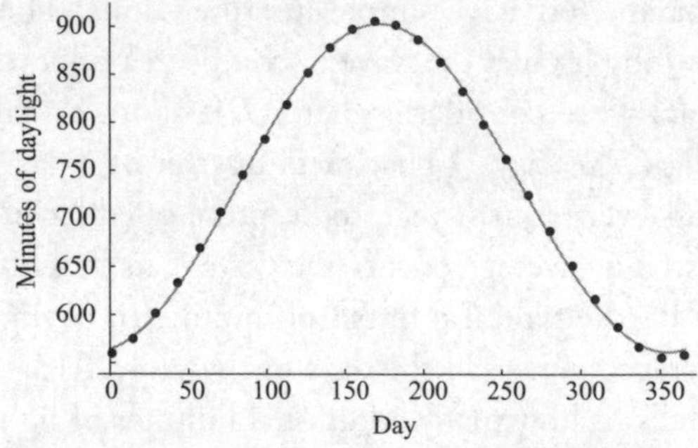

But first, notice the resulting fit is not perfect. For instance, the wave doesn't quite dip down low enough in December, when the days are very short and the data fall below the curve. Nevertheless, a simple sine wave certainly captures the essence of what's going on. Depending on our goals, a fit of this quality may be adequate.

So how does calculus come in? It helps us choose the four parameters optimally. Imagine turning the four knobs to get the best possible fit, somewhat like tuning the dial on a radio to get the strongest possible signal. This is essentially what Fermat did in the overhead-bin problem when he found the roomiest dimensions for the box. In that case, he was tuning a single parameter, *x*, the side length of the box, and looking for a double intersection as a signal that the volume of the box was a maximum. In our case, we have four parameters to tune. But the basic idea is the same. We'll look for a double intersection, and that'll give us our optimal choice of the four parameters.

In a little more detail, here's how it works. For any given choice of the four parameters, we calculate the discrepancy (in other words, the error) between the sine-wave fit and the actual data at every one of the twenty-seven points recorded throughout the year. A natural criterion for choosing the best fit is that the total error, summed over all twenty-seven points, should be as small as we can make it. But total error is not quite the right concept, because we don't want the negative errors to cancel the positive ones and give the false impression that the fit has less error than it does. Undershoots are just as bad as overshoots, and both should be penalized; they shouldn't be allowed to cancel out. For this reason, mathematicians *square* the errors at each point to make the negative ones become positive. That way, they can't possibly produce any spurious cancellations. (Here's one place where the fact that a negative times a negative is a positive is useful in a practical setting. It makes the square of a negative error count as a positive discrepancy, as it should.) So the basic idea is to choose the four parameters in the sine wave in such a way that they minimize the total squared error of the fit to the data. Accordingly, this approach is called the method of least squares. It works best when the data follow a pattern, as they do here.

All of which raises an extremely important general point: Patterns are what make compression possible in the first place. Only patterned data can be compressed. Random data cannot. Happily, many of the things people care about, like songs and faces and fingerprints, are highly structured and patterned. Just as day length follows a simple wave pattern, a photograph of a face contains eyebrows, blemishes, cheekbones, and other characteristic patterns. Songs have melodies and harmonies, rhythms and dynamics. Fingerprints contain ridges and loops and whorls. As human beings, we recognize these patterns instantly. Computers can be taught to recognize them too. The trick is to find the right kinds of mathematical objects to encode particular patterns. Sine waves are ideal for representing periodic patterns, but they are less well suited to representing sharply localized features, like the edge of a nostril or a beauty mark.

For this purpose, researchers in several different fields came up with a generalization of sine waves called wavelets. These little waves are more localized than sine waves. Instead of extending periodically out to infinity in both directions, they are sharply concentrated in time or space.

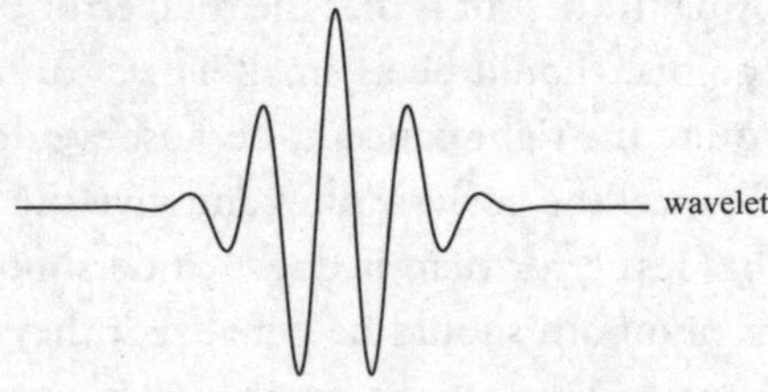

Wavelets suddenly turn on, oscillate for a while, and then turn off. They look almost like the signals on heart monitors or the bursts of activity recorded on seismographs during an earthquake. They are ideal for representing a sudden spike in a brain-wave recording, a bold stroke on a Van Gogh painting, or a wrinkle on a face.

The Federal Bureau of Investigation used wavelets to modernize their fingerprint files. From the time fingerprints were introduced, at the beginning of the twentieth century, fingerprint records had been stored as inked impressions on paper cards. The files were difficult to search quickly. By the mid-1990s, the collection had grown to roughly two hundred million fingerprint cards on file and occupied an acre of office space. When the FBI decided to digitize the files, they turned them into grayscale images with 256 different levels of gray at a resolution of 500 dots per inch, enough to capture all the fine whorls, loops, ridge endings, bifurcations, and other identifying minutiae of fingerprints.

The problem, however, was that at the time, a single digitized card contained about 10 megabytes of data. That made it prohibitive for the FBI to send digital files quickly to local police. Remember, this was in the mid-1990s, when phone modems and fax machines were state of the art and transmitting a 10-megabyte file took hours. Plus it was tough to exchange files that big when 1.5-megabyte floppy disks were the medium of choice. Given the growing demand for faster turnaround times on the thirty thousand new fingerprint cards that flooded in every day with urgent requests for background checks, there was a desperate need to modernize the system. The FBI had to find a way to compress the files without distorting them.

Wavelets were ideal for the job. By representing fingerprints as combinations of many wavelets and by turning the knobs on them optimally using calculus, mathematicians from the Los Alamos National Lab teamed up with the FBI to shrink their files by a factor of more than twenty. It was a revolution for forensics. Thanks to Fermat's ideas in their modern form (along with an even greater role for wavelet analysis, computer science, and signal processing), a 10-megabyte file could be compressed to only 500 kilobytes, a manageable size to send over the phone lines. And it could be done without sacrificing fidelity. Human-fingerprint experts nodded their approval. So did computers; the compressed files passed the FBI's automatic identification system with flying colors. It was good news for calculus and bad news for criminals.

The Principle of Least Time

I wonder what Fermat would have thought of this use of his ideas. He himself was never especially interested in applying mathematics to the real world. He was content to do math for its own sake.

But he did make one contribution to applied mathematics of lasting importance: he was the first person to deduce a law of nature from a deeper law by using calculus as a logical engine. Just as Maxwell would do with electricity and magnetism two centuries later, Fermat translated a hypothetical law of nature into the language of calculus, started the engine, and fed the law in, and out popped another law, a consequence of the first one. In so doing, Fermat, the accidental scientist, initiated a style of reasoning that has dominated theoretical science ever since.

The story began in 1637 when a group of Parisian mathematicians asked Fermat for his opinion on Descartes's recent treatise on optics. Descartes had a theory about how light bent when it passed from air into water or from air into glass, an effect known as refraction.

Anyone who has ever played with a magnifying glass knows that light can be bent and focused. In my youth, I liked to set leaves on fire by holding a magnifying glass over them on the driveway and lifting the glass up and down until the sun's rays focused into a tight white spot of blazing intensity, causing the leaf to smolder and eventually ignite. Refraction is used in less spectacular ways in our spectacles. The lenses in our eyeglasses bend and focus the light rays where they belong, at the right place on the retina to correct faulty vision.

The bending of light also explains an illusion you may have noticed while lounging by a swimming pool on a sunny day. Suppose that at the bottom of the pool there happens to be a shiny, tragically misplaced object like a piece of jewelry.

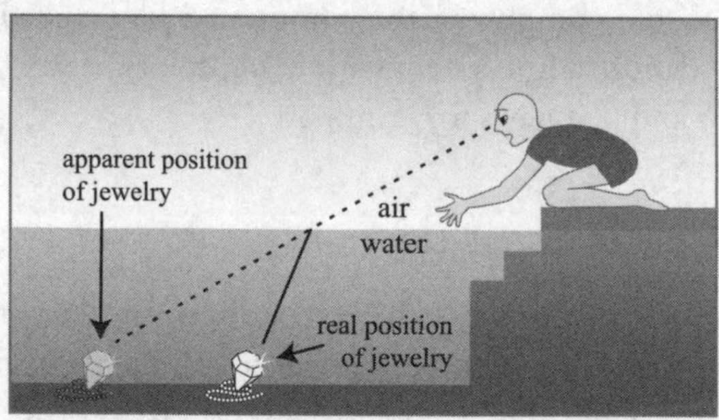

You look down through the water at the shiny object, but it's not quite where it appears to be because the light rays bouncing off it get bent as they pass from water into air on their way out of the pool. For the same reason, a spear fisherman needs to aim *below* the apparent position of a fish to have a chance of hitting it.

Refraction phenomena like these obey a simple rule. When a ray of light passes from a thinner medium like air into a denser medium like water or glass, the ray bends *toward* the perpendicular to the interface between the two media. When it passes from a denser medium into a thinner one, it bends *away* from the perpendicular, as illustrated here.

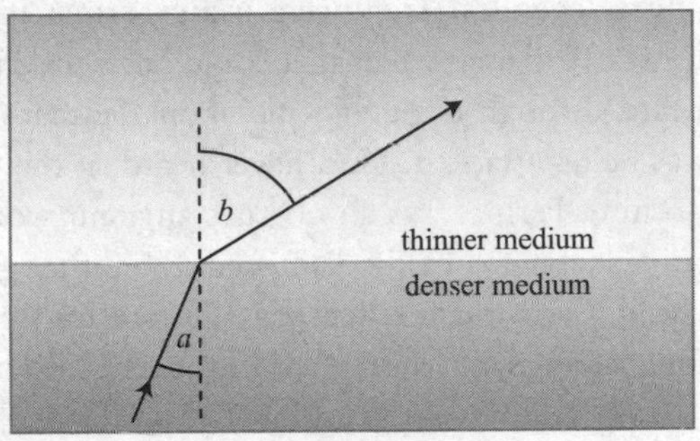

In 1621, the Dutch scientist Willebrord Snell sharpened this rule and made it quantitative by doing a clever experiment. By systematically changing the angle a of the incoming ray and observing how the angle b of the outgoing ray changed in response, he discovered that the ratio $\sin a/\sin b$ always stayed the same for a given pair of media. (Here *sin* refers to the sine function of trigonometry, the same sine function whose wavy graph appeared in our analysis of day length.)

However, Snell found that the value of $\sin a/\sin b$ did depend on what the two media were made of. Air and water produced one constant ratio, whereas air and glass produced another. He had no idea why the sine law worked. It just did. It was a brute fact about light.

Descartes rediscovered Snell's sine law and published it in his 1637 essay *Dioptrics*, unaware that it had been found by at least three others before him: Snell in 1621, the English astronomer Thomas Harriot in 1602, and the Persian mathematician Abu Sa'd al-A'la Ibn Sahl way back in 984.

Descartes had given a mechanical explanation for the sine law in which he (incorrectly) assumed that light moved *faster* in a denser medium. To Fermat, that sounded upside down and contrary to common sense. Trying to be helpful, and being a naive and innocent fellow, Fermat offered what he thought were a few gentle criticisms of Descartes's theory and mailed them back to the Parisian mathematicians who'd asked for his opinion.

What Fermat did not know was that those mathematicians were Descartes's bitter enemies. They were using Fermat for their own sinister purposes. And as any teenager could have anticipated, when Descartes heard through the grapevine about Fermat's comments, he felt he was being attacked. He'd never heard of the lawyer from Toulouse. To him, Fermat was an obscure amateur working out in the countryside, someone easily dismissed as another gnat buzzing around his head. Over the next few years, Descartes treated Fermat condescendingly and claimed that he'd blundered into his results by accident.

Fast-forward twenty years. In 1657, after Descartes had died, Fermat was asked by a colleague named Marin Cureau de la Chambre to revisit the old controversy about refraction. Cureau's inquiry prompted Fermat to take a look at the problem himself, using what he knew about optimization.

Fermat had a hunch that light optimized. More precisely, he guessed that light always followed the path of least resistance between any two points, which he took to mean that it traveled along the fastest possible route. He could see that this *principle of least time* would explain why light moved in a straight line in a uniform medium and why, when it reflected off a mirror, its angle of incidence equaled its angle of reflection. But could the principle of least time also correctly predict how light bent when it passed from one medium into another? Would it explain the sine law of refraction?

Fermat wasn't sure. The calculation wouldn't be easy. An infinite number of straight-line paths, each bent like an elbow at the interface, could take the light from a source point in one medium to a target point in the other.

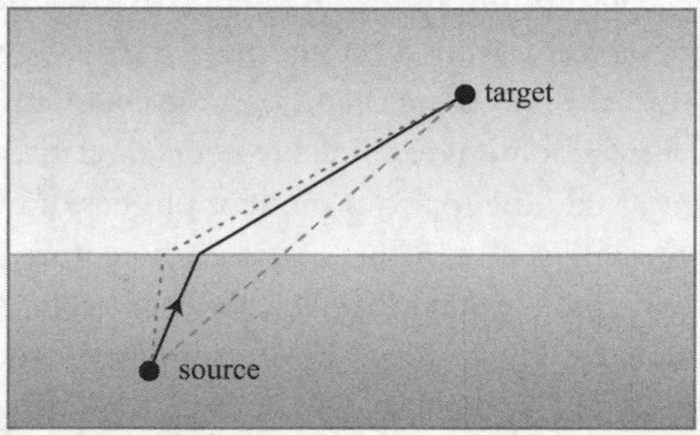

Computing the minimum of all those travel times was going to be difficult, especially at that embryonic stage in the development of differential calculus. There were no tools available other than his old double-intersection method. Plus he was afraid of getting the wrong answer. As he wrote to Cureau, "The fear of finding, after a long and difficult calculation, some irregular and fantastic proportion, and my natural inclination to laziness, left the matter in that state."

Five years passed while Fermat worked on other problems. But eventually his curiosity got the better of him. In 1662 he forced himself to do the calculation. It was arduous and unpleasant. But as he hacked away at the thicket of symbols, he started to see something. The terms began to cancel. The algebra was working. And there it was: the sine law. In a letter to Cureau, Fermat called this calculation "the most extraordinary, the most unforeseen and the happiest" one he'd ever done. "I was so surprised at such an unexpected event, that I can scarcely recover from my astonishment."

Fermat had applied his embryonic version of differential calculus to physics. No one had ever done that before. And in so doing, he showed that light travels in the most efficient way—not the most direct way, but the fastest. Of all the possible paths light can take, it somehow knows, or behaves as if it knows, how to get from here to there as quickly as possible. This was an important early clue that calculus was somehow built into the operating system of the universe.

The principle of least time was later generalized to the principle of least action, where action has a technical meaning that we needn't go into here. This optimization principle—that nature behaves in the most economical way, in a certain precise sense—was found to correctly predict the laws of mechanics. In the twentieth century, the principle of least action was extended to general relativity and quantum mechanics and other parts of modern physics. It even made an impression on philosophy in the seventeenth century, when Gottfried Wilhelm Leibniz argued that all is for the best in the best of all possible worlds, an optimistic point of view later parodied by Voltaire in *Candide*. The idea of using an optimality principle to explain physical phenomena and to deduce its consequences with calculus began with this very calculation by Fermat.

The Tussle over Tangents

Fermat's optimization techniques also allowed him to figure out tangent lines to curves. This was the problem that really made Descartes's blood boil.

The word *tangent* comes from a Latin root for "touching." The terminology is apt, since instead of cutting across a curve in two places, a tangent line touches the curve at one point, barely grazing it.

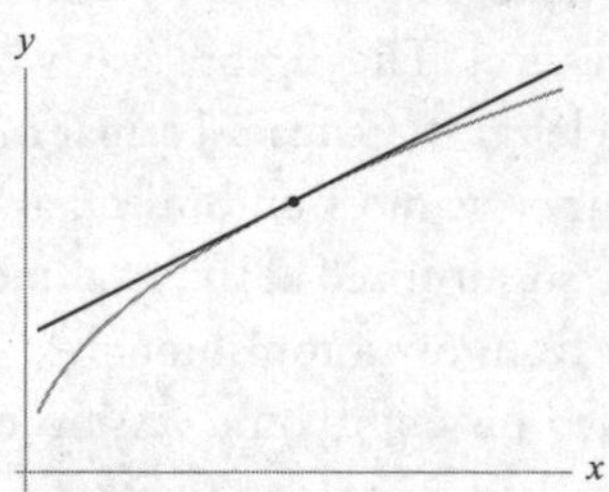

The condition for tangency is similar to that for a maximum or a minimum. If we intersect a curve with a line and then slide the line up or down continuously, tangency occurs when two intersections coalesce into one.

By sometime in the late 1620s, Fermat was able to find the tangent line to essentially any algebraic curve (meaning a curve expressible solely in terms of whole number powers of x and y, without any logarithms, sine functions, or other so-called transcendental functions in it). Using his big idea of the double intersection, he could calculate everything with his methods that we can today with derivatives.

Descartes had his own method of finding tangent lines. In his *Geometry* of 1637, he proudly announced his method to the world. Unaware that Fermat had already solved the problem, Descartes independently hit on the double-intersection idea, but he used circles instead of lines to cut through the curves of interest. Near the point of tangency, a typical circle would cut through the curve at two points, or at none.

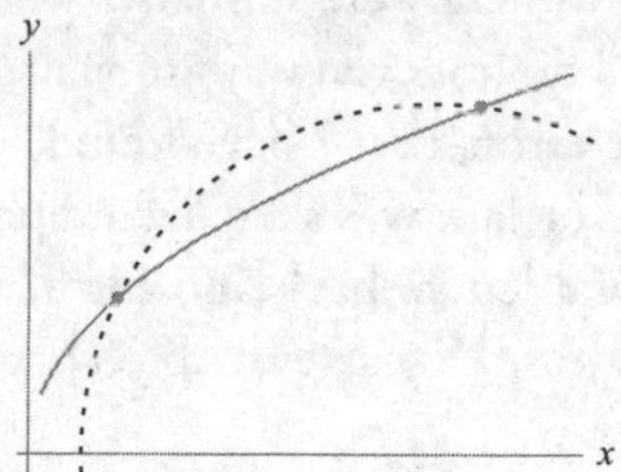

By adjusting the location and radius of the circle, Descartes could force the two intersection points to merge into one. At that double intersection—bingo!—the circle intersected the curve tangentially.

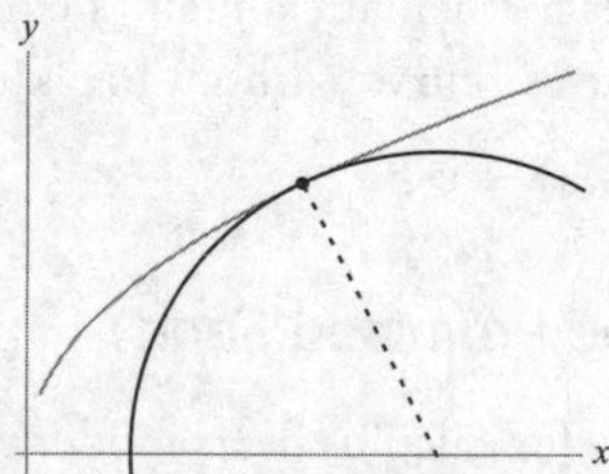

That gave Descartes everything he needed to find the tangent to the curve. It also gave him the normal to the curve, which lay at right angles to the tangent, along the circle's radius.

His method was correct but clumsy. It generated bushels of algebra, much more than Fermat's. But Descartes hadn't even heard of Fermat yet, so in his usual cocksure fashion, he presumed he had outdone everyone. As he crowed in *Geometry*, "I have given a general method of drawing a straight line making right angles with the curve at an arbitrarily chosen point upon it. And I dare say that this is not only the most useful and most general problem in geometry that I know, but even that I ever desired to know."

Late in 1637, when Descartes learned from his correspondents in Paris that Fermat had beaten him to the solution of the tangent problem by about ten years but had never gotten around to publishing it, he was dismayed. In 1638 he studied Fermat's method, looking for holes. Oh, there were so many! Writing through an intermediary, he said, "I do not even want to name him, so that he will feel less shame at the errors that I have found." He challenged Fermat's logic, which, to be fair, was sketchy and poorly explained. But eventually, after several letters back and forth, with Fermat calmly trying to clarify his ideas, Descartes had to concede that his reasoning was valid.

But before admitting defeat, he tried to stump Fermat by challenging him to find the tangent line to a curve defined by the cubic equation $x^3 + y^3 = 3axy$, where a was a constant. Descartes knew that he himself couldn't find the tangent with his own clunky circle method—the algebra became unmanageable—so he was confident that Fermat wouldn't be able to find it with his line method. But Fermat was a stronger mathematician, and he had a better method. He dispatched Descartes's curve without breaking a sweat, much to Descartes's chagrin.

Within Sight of the Promised Land

Fermat paved the way for calculus in its modern form. His principle of least time revealed that optimization is woven deeply into the fabric of nature. His work on analytic geometry and tangent lines blazed a trail to differential calculus that others soon followed. And his virtuosity with algebra enabled him to find the areas under certain curves that had eluded even his most illustrious predecessors. In particular, he found the area under the curve $y = x^n$ for any positive integer n, using little more than his bare hands. (Others had solved the first nine cases, $n = 1, 2, \ldots, 9$, but couldn't find a strategy that worked for all n.) Fermat's advance was a giant step forward for integral calculus, one that would set the stage for breakthroughs to come.

Yet for all that, his studies still fell short of the secret that Newton and Leibniz would soon discover, the secret that revolutionized and unified the two sides of calculus. It's a pity that Fermat missed it, for he came so close. The missing link was related to something he created but never recognized as crucial, something implicit in his method of maxima and tangents. It would later be called the derivative. Its applications would go far beyond curves and their tangents to include any sort of change at all.

5

The Crossroads

We have come to a crossroads in our story. This is where calculus becomes modern and progresses from the mystery of curves to the mysteries of motion and change. It's where calculus starts to wonder about the rhythms of the universe, its ups and downs, its ineffable patterns in time. No longer content in the static world of geometry, calculus becomes fascinated with dynamics. It asks: What are the rules of motion and change? What can we predict about the future with certainty?

In the four centuries since calculus reached this crossroads, it has branched out from algebra and geometry to physics and astronomy, biology and medicine, engineering and technology, and every other field where all is in flux and change never stops. Calculus has mathematized time. And it has offered us hope that the world we live in, for all its unfairness and misery and chaos, may be reasonable deep down, deep in its heart, where it follows mathematical laws. Sometimes we can find those laws through science. Sometimes we can understand them through calculus. And sometimes we can use them to improve our lives, help our societies, and change the course of history for the better.

The pivotal moment in the story of calculus occurred in the middle of the seventeenth century when the mysteries of curves, motion, and change collided on a two-dimensional grid, the *xy* plane of Fermat and Descartes. Back then, Fermat and Descartes had no idea what a versatile tool they'd created. They intended the *xy* plane as a tool for pure mathematics. Yet from the start, it too was a crossroads of sorts, a place where equations met curves, algebra met geometry, and the mathematics of the East met that of the West. Then, in the next generation, Isaac Newton built on their work as well as on the work of Galileo and Kepler and brought geometry and physics together in a great synthesis. Newton's spark set off the fire that lit the Enlightenment and caused a revolution in Western science and mathematics.

But to tell that story, we need to begin with the arena where it all took place, the *xy* plane. When students today take their first course in calculus, they spend the entire year in that plane. The term of art for this subject is *calculus of functions of one variable*. Our discussion of it will occupy us for the next several chapters. We begin here with functions.

In the centuries since curves collided with motion and change, the *xy* plane has become ever more vital as a hub. It's used today in every quantitative field to graph data and uncover hidden relationships. We can use it to visualize how one variable depends on another, how x relates to y when everything else is held constant. Such relationships are modeled by functions of one variable. They are written symbolically as $y = f(x)$, which is pronounced "y equals f of x." Here f denotes a function that describes how the variable y (called the *dependent* variable) depends on the variable x (the *independent* variable), assuming everything else is nailed down and held constant. Such functions model how the world behaves at its tidiest. A cause produces a predictable effect. A dose stimulates a predictable response. More formally, a function f is a rule that assigns a unique y to each x. It's like an input-output machine: feed it x and it spits out y, and it does so reliably and predictably.

A few decades before Fermat and Descartes, Galileo understood the power of this deliberate simplification of reality. He meticulously changed just one thing at a time in his experiments while holding everything else constant. He let a ball roll down a ramp and measured how far it went in a certain amount of time. Nice and simple —distance as a function of time. Likewise, Kepler studied how long it took a planet to orbit the sun and related that period to the planet's average distance from the sun. One variable versus another, period versus distance. This was the way to make progress. This was the way to read the great book of nature.

We've encountered examples of functions in previous chapters. In the cinnamon-raisin bread example, x was the number of slices eaten and y was the number of calories consumed. The relationship in that case was $y = 200x$, which produced a straight-line graph in the xy plane. Another example came up when we studied how the length of the day changed with the seasons in New York City in 2018. In that setting, the variable x represented the day of the year and y was the number of minutes of daylight on that day, defined as the time from sunrise to sunset. We found that the graph in that case oscillated like a sine wave, with the longest days in the summer and the shortest in the winter.

The Function of Functions

Some functions are so important that they've been given their own buttons on a scientific calculator. These are mathematical celebrities like x^2 and $\log x$ and 10^x. Admittedly, most people don't have much use for them. They aren't needed for making change or deciding how much to tip. In everyday life, numbers are usually enough. That's why when you press the calculator app on your phone, by default it offers you a basic calculator with the numbers from 0 to 9 on it, as well as the four basic operations of arithmetic—addition, subtraction, multiplication and division—and a button for percentages. Those are all most of us need as we go about our business.

But for people in technical professions, numbers are just the beginning. Scientists, engineers, financial quants, and medical researchers need to work with *relationships* between numbers, which show how one thing affects another. To describe relationships like that, functions are indispensable. They provide the tools needed to model motion and change.

Generally speaking, things can change in one of three ways: they can go up, they can go down, or they can go up and down. In other words, they can grow, decay, or fluctuate. Different functions are suitable for different occasions. Since we're going to be meeting various functions in the pages ahead, it's helpful to recall some of the most useful ones.

Power Functions

To quantify growth in its most gradual forms, we often use *power functions* like x^2 or x^3, in which a variable x is raised to some power.

The simplest of these is a linear function, in which the dependent variable y grows in direct proportion to x. For example, if y is the number of calories consumed by eating 1, 2, or 3 slices of cinnamon-raisin bread, then y grows according to the equation $y = 200x$, where x is the number of slices and 200 is the number of calories per slice. However, there is no need for a separate x button on the calculator because multiplication serves the same purpose; here, 200 calories times the number of slices of bread equals the number of calories consumed.

But for the next kind of growth in the hierarchy, the type known as quadratic growth, it's very helpful to have an x^2 button on the calculator. Quadratic growth is less intuitive than linear growth. It's not just a matter of multiplication. For example, if we change x from 1 to 2 to 3 again and ask how the corresponding values of $y = x^2$ change, we see they go from $1^2 = 1$ to $2^2 = 4$ to $3^2 = 9$. Thus the y-values grow in increasing steps, first by $\Delta y = 4 - 1 = 3$, then by $\Delta y = 9 - 4 = 5$. If we keep going, they'd increase by 7, 9, 11, and so on, following the pattern of odd numbers. Thus, for quadratic growth, the amount of change itself goes up as we increase x. The growth grows faster as it proceeds.

We already saw this curious odd-number pattern in Galileo's inclined-plane experiments in which he timed balls as they rolled slowly down a ramp. He observed that when a ball was released from rest, it rolled faster as time passed, such that in each successive increment of time, it traveled farther and farther, with the successive distances growing in proportion to the successive odd numbers 1, 3, 5, and so on. Galileo realized what this cryptic rule implied. It meant that the total distance the ball rolled wasn't proportional to time; it was proportional to time *squared*. So in the study of motion, the squaring function x^2 arose very naturally.

Exponential Functions

In contrast to a mild power function like x or x^2, an exponential function like 2^x or 10^x describes a much more explosive kind of growth, a growth that snowballs and feeds on itself. Instead of *adding* a constant increment at each step as in linear growth, exponential growth involves *multiplying* by a constant factor.

For example, a bacterial population growing on a petri dish doubles every 20 minutes. If there are 1000 bacterial cells initially, after 20 minutes, there will be 2000 cells. After another 20 minutes, 4000 cells, and 20 minutes after that, 8000 cells, then 16,000, 32,000, and so on. In this example, the exponential function 2^x comes into play. Specifically, if we measure time in units of 20 minutes, the number of bacteria after x units of time would be 1000×2^x cells. Similar exponential growth is relevant to all sorts of snowballing processes, from the multiplication of real viruses to the viral spread of information in a social network.

Exponential growth is also relevant to the growth of money. Imagine a lump sum of \$100 sitting in a bank account that earns a constant annual interest rate of 1 percent. After one year, that sum would grow to \$101. After two years, it would become \$101 times 1.01, which equals \$102.01. After x years, the amount of money in the bank account would be $100 \times (1.01)^x$.

In exponential functions like 2^x and $(1.01)^x$, the numbers 2 and 1.01 are called the function's base. The most commonly used base in precalculus mathematics is 10. There's no mathematical reason for preferring 10 over any other base. It's a traditional favorite because of an accident of biological evolution: we happen to have ten fingers. Accordingly, we have based our system of arithmetic, the decimal system, on powers of ten.

For the same reason, the exponential function that all budding scientists encounter first, usually in high school, is 10^x. Here the number x is called the exponent. When x is 1, 2, 3, or any other positive whole number, that value of x indicates how many factors of ten are being multiplied together in 10^x. But when x is zero, negative, or in between two whole numbers, the meaning of 10^x is a bit subtler, as we're about to see.

Powers of Ten

There are many situations in science where we use powers of ten to ease calculations. In particular, when numbers are either very big or very small, rewriting them in scientific notation is a good idea. Scientific notation uses powers of ten to express numbers as compactly as possible.

Take the number twenty-one trillion, much talked about these days in connection with the US national debt. Twenty-one trillion can be written either in decimal notation as 21,000,000,000,000, or more compactly in scientific notation as $21 \times 10^{12} = 2.1 \times 10^{13}$. If for some reason we need to multiply that big number by, say, one billion, it's easier to write $(2.1 \times 10^{13}) \times 10^9 = 2.1 \times 10^{22}$ than it is to keep track of all those zeros in decimal notation.

The first three powers of ten are numbers we run into every day:

1 $10^1 = 10$
2 $10^2 = 100$
3 $10^3 = 1000$

Notice the trend: The left column (x) grows *additively*, whereas the right column (10^x) grows *multiplicatively*, as we expect for exponential growth. Thus, in the left column, each upward step adds 1 to the preceding number, while in the right column it multiplies the preceding number by 10. This intriguing correspondence between addition and multiplication is a hallmark of exponential functions in general and powers of ten in particular.

Because of this correspondence between the two columns, if we *add* two numbers in the left column, that operation corresponds to *multiplying* their partners in the right column. For example, 1 + 2 = 3 on the left translates into $10 \times 100 = 1000$ on the right. The translation from addition to multiplication makes sense because

$$10^{1+2} = 10^3 = 10^1 \times 10^2.$$

Thus, when we multiply powers of ten, their exponents add, as 1 and 2 do here. The general rule is

$$10^a \times 10^b = 10^{a+b}.$$

A related trend is that *subtraction* in the left column corresponds to *division* in the right column:

$$3 - 2 = 1 \text{ corresponds to } \tfrac{1000}{100} = 10.$$

These nifty patterns suggest how to continue the two columns downward toward smaller and smaller numbers. The principle is, whenever we subtract by 1 in the left column, we should divide by 10 in the right column. Now look at the top row again:

1 $10^1 = 10$
2 $10^2 = 100$
3 $10^3 = 1000$

Since subtracting 1 on the left amounts to dividing by 10 on the right, the correspondence continues with a new top row that has $1 - 1 = 0$ on the left and $10/10 = 1$ on the right:

0 $10^0 = 1$
1 $10^1 = 10$
2 $10^2 = 100$
3 $10^3 = 1000$

This reasoning explains why 10^0 is defined as 1 (and has to be defined that way), a definition that many people find puzzling. Any other choice would break the pattern. It's the only definition that continues the trends established farther down in the two columns.

Going on in this way, we can extrapolate the correspondence even further, now to negative numbers in the left column. The corresponding numbers on the right then become fractions, equivalent to powers of $\frac{1}{10}$:

–2	$\frac{1}{100}$
–1	$\frac{1}{10}$
0	1
1	10
2	100
3	1000

Notice that the numbers in the right column always remain positive, even when the numbers in the left column become zero or negative.

A potential cognitive pitfall when using powers of ten is that they can make vastly different numbers seem more similar than they really are. To avoid this trap, it's good to pretend that different powers of ten form conceptually distinct categories. Sometimes human languages do this on their own by assigning distinct names to different powers of ten, as if they were unrelated species. In English we refer to 10, 100, and 1000 with three unrelated words—*ten*, a *hundred*, and a *thousand*. That's good. It conveys the right idea that these numbers are qualitatively different, even though they are neighboring powers of ten. Anyone who appreciates the difference between a five-figure and a six-figure salary knows that one extra zero can matter a great deal.

When the words for powers of ten sound too much alike, we are led astray. During the 2016 US presidential campaign, Senator Bernie Sanders frequently railed against the exorbitant tax breaks going to "millionaires and billionaires." Whether you agreed with him or not about the politics, he unfortunately made it sound like, in terms of wealth, millionaires and billionaires were comparable. In fact, billionaires are much, much richer than millionaires. To grasp how different a million is from a billion, think about it like this: A million seconds is a little under two weeks; a billion seconds is about thirty-two years. The first is the length of a vacation; the second is a significant fraction of a lifetime.

The lesson here is that we need to use powers of ten with care. They are dangerously strong compressors, capable of shrinking enormous numbers down to sizes we can fathom more easily. That's also why they're so popular with scientists. In contexts in which some quantity varies over many orders of magnitude, powers of ten are often used to define an appropriate measurement scale. Examples include the pH scale of acidity and basicity, the Richter scale of earthquake magnitudes, and the decibel measure of loudness. For instance, if the pH of a solution changes from 7 (neutral, like pure water) to 2 (acidic, like lemon juice), the concentration of hydrogen ions increases by five orders of magnitude, meaning a factor of 10^5, or a hundred thousand. The drop in pH from 7 to 2 makes it seem like just five itty-bitty steps, not much of a change at all, even though it's really a hundred-thousand-fold change in hydrogen-ion concentration.

Logarithms

In the examples we have considered so far, the numbers in the right column, like 100 and 1000, have always been round numbers. Since powers of ten are so convenient, it would be wonderful if we could express non-round numbers in the same manner. Take 90, for instance. Given that 90 is a little less than 100, and 100 equals 10^2, it seems like 90 should equal 10 raised to some number slightly less than 2. But raised to what number, exactly?

Logarithms were invented to answer such questions. On a calculator, if you type in 90 and then press the log button, you get

$$\log 90 = 1.9542\ldots.$$

That's the answer: $10^{1.9542\ldots} = 90$.

In this way, logarithms enable us to write any positive number as a power of ten. Doing that makes many calculations easier and also reveals surprising connections between numbers. Look what happens if we multiply 90 by a factor of 10 or 100 and then take its log again:

$$\log 900 \approx 2.9542\ldots$$

and

$$\log 9000 = 3.9542\ldots.$$

Observe two striking things here:

1) All the logs here have the same decimal part, .9542 . . . .
2) Multiplying the original number, 90, by 10 increased its log by 1. Multiplying it by 100 increased its log by 2, etc.

We can explain both of these facts by appealing to a rule of logs: *The log of a product is the sum of the logs.* Here,

$$\begin{aligned}\log 90 &= \log(9 \times 10)\\ &= \log 9 + \log 10\\ &= .9542\ldots\ +1\end{aligned}$$

and

$$\begin{aligned}\log 900 &= \log(9 \times 100)\\ &= \log 9 + \log 100\\ &= .9542\ldots\ +2\end{aligned}$$

and so on. This explains why the logs of 90 and 900 and 9000 all have the same decimal part, .9542 . . . . That decimal part is the log of 9, and 9 is a factor that appears in all of the numbers we've been discussing. The different powers of ten show up as the different whole-number parts in the logs (in this case, 1, 2, or 3 in front of the decimal part). Because of this, if we are interested in the logs of other numbers, we need only work out the logs of numbers from 1 to 10.

That takes care of the decimal parts. The log of every other positive number can then be expressed in terms of those logs alone. Powers of ten have their own job; they account for the whole-number part.

The general rule in symbolic form is

$$\log(a \times b) = \log a + \log b.$$

In other words, when we multiply two numbers together and then take their log, the result is the sum (not the product!) of their individual logs. In that sense, logarithms replace multiplication problems with addition problems, which are *much* easier. This is why logarithms were invented. They sped up calculations tremendously. Instead of having to deal with Herculean multiplication problems, square roots, cube roots, and the like, such calculations could be turned into addition problems and then solved with the help of a lookup table known as a table of logarithms. The idea of logarithms was in the air in the early seventeenth century, but much of the credit for popularizing them goes to the Scottish mathematician John Napier, who published his *Description of the Wonderful Rule of Logarithms* in 1614. A decade later, Johannes Kepler enthusiastically used the new calculational tool when he was compiling astronomical tables about the positions of the planets and other heavenly bodies. Logarithms were the supercomputers of their era.

Many people find logarithms confusing, but they make a lot of sense if you think about them by analogy with carpentry. Logarithms and other functions are like tools. Different tools have different purposes. Hammers are for pounding nails into wood; drills are for boring holes; saws are for cutting. Likewise, exponential functions are for modeling growth that feeds on itself, and power functions are for modeling less violent forms of growth. Logarithms are useful for the same reason that staple removers are useful: they undo the action of another tool. Specifically, logarithms undo the actions of exponential functions, and vice versa.

Consider the exponential function 10^x and apply it to a number, say 3. The result is 1000. To undo that action, press the log x button. Applying it to 1000 returns the number we started with: 3.

The base-10 logarithm function log x undoes the action of the 10^x function. They are inverse functions in this sense.

Aside from their role as inverse functions, logarithms also describe many natural phenomena. For example, our perception of pitch is approximately logarithmic. When a musical pitch goes up by successive octaves, from one *do* to the next, that increase corresponds to successive doublings of the frequency of the associated sound waves. Yet although the waves oscillate twice as fast for every octave increase, we hear the doublings—which are *multiplicative* changes in frequency—as equal upward steps in pitch, meaning equal *additive* steps. It's freaky. Our minds fool us into believing that 1 is as far from 2 as 2 is from 4, and as 4 is from 8, and so on. We somehow sense frequency logarithmically.

The Natural Logarithm and Its Exponential Function

As useful as base 10 was in its heyday, it is rarely deployed in modern calculus. It has been superseded by another base that looks abstruse but turns out to be far more natural than 10. This natural base is called e. It's a number close to 2.718 (I'll explain where it comes from in a minute), but its numerical value is beside the point. The important point about e is that an exponential function with this base grows at a rate precisely equal to the function itself.

Let me say that again.

The rate of growth of e^x is e^x itself.

This marvelous property simplifies all calculations about exponential functions when they are expressed in base e. No other base enjoys this simplicity. Whether we are working with derivatives, integrals, differential equations, or any of the other tools of calculus, exponential functions expressed in base e are always the cleanest, most elegant, and most beautiful.

Aside from its simplifying role in calculus, base e arises naturally in finance and banking. The following example will reveal where the number e comes from and how it is defined.

Imagine you deposit $100 in a bank that pays interest at an

implausible but irresistible annual rate of 100 percent. That means that after one year, your \$100 would become \$200. Now start over and consider an even more favorable scenario. Imagine that you could persuade the bank to compound your money twice a year so that you could gain interest on the interest as your money grows. How much more would you earn in that case? Since you're asking the bank to compound the money twice as often, it's only fair that the interest rate for each six-month period should be half as large, namely 50 percent. Thus, after six months, you'd have $\$100 \times 1.50$, which equals \$150. Six months later, at year's end, the amount would be another 50 percent more: $\$150 \times 1.50$, which equals \$225. That's more than the \$200 you got under the original arrangement because you gained interest on the interest during the year.

The next question is, what happens if you could get the bank to compound your money more and more frequently, at correspondingly smaller interest rates during each compounding period? Would you achieve fabulous wealth? Unfortunately, no. Compounding quarterly would yield $\$100 \times (1.25)^4 \approx \$244.14$, not much of an improvement over \$225. Compounding still more rapidly, once a day for the 365 days in the year, would leave you with only

$$\$100 \times (1+\tfrac{1}{365})^{365} \approx \$271.46$$

at the end of the year. Here the 365 in both the denominator and the exponent refers to the number of compounding periods in the year, and the 1 in the numerator of $1/365$ is the 100 percent interest rate expressed as a decimal.

Finally, suppose we take this compounding madness to the limit. If the bank compounds your money *n* times a year where *n* is a monstrously huge number, with correspondingly tiny interest rates during each sub-nanosecond period, then by analogy with the result for 365 daily periods, you'd have

$$\$100 \times (1+\tfrac{1}{n})^{n}$$

in your account at year's end. As n approaches infinity, this amount approaches 100 times the limit of

$$\left(1+\frac{1}{n}\right)^n$$

as n approaches infinity. That limit is defined as the number e. It's not at all obvious what the limiting number is, but it turns out to be approximately 2.71828 . . . .

In the banking world, the financial arrangement above is called *continuous compounding.* Our results show that it is nothing to get excited about. In the problem above, it would yield a year-end balance of

$$\$100 \times e \approx \$271.83.$$

That's the best deal yet, but it's only 37 cents more than the result of daily compounding.

We just jumped through a lot of hoops to define e. In the end, e turned out to be a complicated limit. It has infinity built into it in much the same way that the number π does for circles. Recall that π involved a calculation of the perimeter of a many-sided polygon inscribed in a circle. That polygon approached the circle as the number of sides, n, approached infinity and the lengths of those sides approached zero. The number e is defined in a somewhat similar way as a limit, except that it arises in the different context of continuously compounded growth.

The exponential function associated with e is written as e^x, just as the exponential function for base 10 is written as 10^x. It looks weird at first, but at a structural level it's just like base 10. All the principles and patterns are the same. For example, if we want to find an x such that e^x is a given number, say 90, we can again use logarithms as we did before, except now we wheel out the base-e logarithm, better known as the *natural logarithm* and denoted $\ln x$. To find the unknown x such that $e^x = 90$, turn on a scientific calculator, enter 90, press the $\ln x$ button, and there's your answer:

$$\ln 90 \approx 4.4998.$$

To check it, keep that number on the screen and hit the e^x button. You should get 90. As before, logs and exponentials undo each other's actions like a stapler and a staple remover.

Recondite as all this may sound, the natural logarithm is extremely practical, though often inconspicuously. For one thing, it underlies a rule of thumb known to investors and bankers as the rule of 72. To estimate how long it will take to double your money at a given annual rate of return, divide 72 by the rate of return. Thus, money growing at a 6 percent annual rate doubles after about $72/6 = 12$ years. This rule of thumb follows from the properties of the natural logarithm and exponential growth and works well if the interest rate is low enough. Natural logarithms also operate behind the scenes in the carbon dating of ancient trees and bones and in art-authentication disputes. A famous case involved paintings allegedly by Vermeer that turned out to be forgeries; this was revealed by analysis of the radioactively decaying isotopes of lead and radium in the paint. As these examples suggest, the natural logarithm now pervades all fields where exponential growth and decay arise.

The Mechanism Behind Exponential Growth and Decay

To reiterate the main point, the thing that makes e special is that the rate of change of e^x is e^x. Hence, as the graph of this exponential function soars higher and higher, its slope always tilts to match its current height. The higher it gets, the steeper it climbs. In the jargon of calculus, e^x is its own derivative. No other function can say that. It's the fairest of them all—at least as far as calculus is concerned.

Although base e is uniquely distinguished, other exponential functions obey a similar principle of growth. The only difference is that the rate of exponential growth is *proportional* to the function's current level, not strictly *equal* to it. Still, that proportionality is sufficient to generate the explosiveness we associate with exponential growth.

The explanation for the proportionality should be intuitively clear. In bacterial growth, for example, bigger populations grow faster because the more cells there are, the more of them are available to divide and make offspring. The same is true with the growth of money in an account being compounded at a constant interest rate. More money means more interest on that money and hence a faster rate of growth of the overall account.

This effect also accounts for the howl of a microphone when it picks up the sound of its own loudspeaker. The loudspeaker contains an amplifier that makes a sound louder. In effect, it multiplies the volume of the sound by a constant factor. If that louder sound gets picked up by the microphone and then sent back through the amplifier again, its volume will be amplified repeatedly in a positive feedback loop. This causes a sudden exponential runaway of volume, growing at a rate proportional to the current volume and leading to the awful screeching sound.

Nuclear chain reactions are governed by exponential growth for the same reason. When a uranium atom splits, it fires out neutrons that can potentially smash into other atoms and cause them to split, sending out still more neutrons, and so on. The exponential growth of the number of neutrons, if left unchecked, can set off a nuclear explosion.

Along with growth, decay can be described by exponential functions. Exponential decay occurs when something is being depleted or consumed at a rate proportional to its current level. For example, half the atoms in an isolated lump of uranium always take the same amount of time to decay radioactively, no matter how many atoms were present in the lump initially. That decay time is known as the half-life. The concept applies to other fields as well. In chapter 8 we'll discuss what doctors learned about AIDS after they discovered that the number of virus particles in the bloodstreams of HIV-infected patients dropped exponentially, with a half-life of only two days, after a miracle drug called a protease inhibitor was administered.

These diverse examples, from the dynamics of chain reactions and microphone feedback howl to the accumulation of money in a bank account, make it seem like exponential functions and their logarithms are firmly planted in the part of calculus that deals with changes in time. And it's true that exponential growth and decay are prominent topics on the modern side of the crossroads of calculus. But logarithms were first sighted on the other side, back when calculus was still focused on the geometry of curves. Indeed, the natural logarithm arose early on in studies of the area under the hyperbola $y = 1/x$. The plot thickened in the 1640s when it was discovered that the area under the hyperbola defined a function that behaved uncannily like a logarithm. In fact, it *was* a logarithm. It obeyed the same structural rules and turned problems of multiplication into problems of addition, just like any other logarithm, but its base was unknown.

There was still much to be learned about the areas under curves. That was to be one of the two great challenges ahead for calculus. The other was to devise a more systematic method to find the tangent lines and slopes of curves. The solution of these two problems and the discovery of the surprising connection between them would soon take calculus, and the world, decisively into modernity.

6

The Vocabulary of Change

FROM A TWENTY-FIRST-CENTURY vantage point, calculus is often seen as the mathematics of change. It quantifies change using two big concepts: derivatives and integrals. Derivatives model rates of change and are the main topic of this chapter. Integrals model the accumulation of change and will be discussed in chapters 7 and 8.

Derivatives answer questions like "How fast?" "How steep?" and "How sensitive?" These are all questions about *rates of change* in one form or another. A rate of change means a change in a dependent variable divided by a change in an independent variable. In symbols, a rate of change always takes the form $\Delta y/\Delta x$, a change in y divided by a change in x. Sometimes other letters are used, but the structure is the same. For example, when time is the independent variable, it's customary and clearer to write the rate of change as $\Delta y/\Delta t$, where t denotes time.

The most familiar example of a rate is a *speed*. When we say a car is going 100 kilometers an hour, that number qualifies as a rate of change because it defines speed as a $\Delta y/\Delta t$ when it states how far the car goes (Δy = 100 kilometers) in a given amount of time (Δt = 1 hour).

Likewise, *acceleration* is a rate. It's defined as the rate of change of speed, usually written $\Delta v/\Delta t$, where v stands for velocity. When the American car manufacturer Chevrolet claims that one of its muscle cars, the V-8 Camaro SS, can go from 0 to 60 miles per hour in 4 seconds flat, they're quoting acceleration as a rate: a change in speed (from 0 to 60 miles per hour) divided by a change in time (4 seconds).

The *slope* of a ramp is a third example of a rate of change. It's defined as the ramp's vertical rise Δy divided by its horizontal run Δx. A steep ramp has a large slope. A wheelchair-accessible ramp is required by US law to have a slope less than 1⁄12. Flat ground has zero slope.

Of all the various rates of change that exist, the slope of a curve in the *xy* plane is the most important and useful, because it can stand in for all the rest. Depending on what *x* and *y* represent, the slope of a curve can indicate a speed, an acceleration, a rate of pay, an exchange rate, the marginal return on an investment, or any other kind of rate. For example, when we plotted the number of calories, *y*, contained in *x* slices of cinnamon-raisin bread, the graph was a line with a slope of 200 calories per slice. That slope, a geometrical feature, told us the rate at which the bread delivers calories, a nutritional feature. Similarly, on a graph of distance versus time for a moving car, the slope indicates the car's speed. Thus, slope is a sort of universal rate. Since any function of one variable can always be graphed as a curve on the *xy* plane, we can find its rate of change by reading off the slope of its graph.

The catch is that rates of change are rarely constant in the real world or in mathematics. In that case, defining a rate becomes problematic. The first big issue in differential calculus is to define what we mean by the rate when the rate of change keeps changing. Speedometers and GPS devices have solved this problem. They always know what speed to report even if a car speeds up and slows down. How do these gadgets do it? What calculation are they making? With calculus, we'll see.

Just as speeds don't need to be constant, slopes don't need to be constant either. On a curve like a circle or a parabola or any other smooth path (as long as it's not a perfectly straight line), the slope is bound to be steeper in some places and shallower in others. That's true in the real world too. Mountain trails have treacherous steep sections and restful flat sections. So the question remains: How do we define the slope when the slope keeps changing?

The first thing to realize is that we need to expand our concept of what a rate of change is. In algebra problems that involve distance equals rate times time, the rate is always a constant. That is not the case in calculus. Because speeds, slopes, and other rates vary as the independent variable x or t changes, they have to be regarded as functions themselves. Rates of change can no longer be mere numbers. They need to become functions.

This is what the concept of a *derivative* does for us. It defines a rate of change as a function. It specifies a rate at a given point or at a given time, even if that rate is variable. In this chapter, we'll see how derivatives are defined, what they mean, and why they matter.

To let the cat out of the bag, derivatives matter because they're ubiquitous. At their deepest level, the laws of nature are expressed in terms of derivatives. It's as if the universe knew about rates of change before we did. At a more mundane level, derivatives come up whenever we want to quantify how a change in something is related to a change in something else. How much does raising the price of an app affect the consumer demand for it? How much does increasing the dose of a statin drug enhance its ability to lower a patient's cholesterol or increase its risk of triggering side effects like liver damage? Whenever we study a relationship of any kind, we want to know: If one variable changes, how much does a related variable change? And in what direction, up or down? These are questions about derivatives. The acceleration of a rocket ship, the growth rate of a population, the marginal return on an investment, the temperature gradient in a bowl of soup—derivatives, one and all.

In calculus, the symbol for the derivative is dy/dx. It's supposed to remind you of an ordinary rate of change $\Delta y/\Delta x$, except that the two changes dy and dx are now imagined to be infinitesimally tiny. That's a wild new idea that we'll work our way up to, slowly and gently, though it shouldn't come as a surprise. We know from the Infinity Principle that the way to make progress on complicated problems is to chop them into infinitesimal bits, analyze the bits, and then put the bits back together to find the answer. The little changes *dx* and *dy* are those infinitesimal bits in the context of differential calculus. Putting them back together is the job of integral calculus.

The Three Central Problems of Calculus

To prepare ourselves for what lies ahead, we need to have the big picture in mind from the start. There are three central problems in calculus. They are shown schematically on the diagram below.

1) *The forward problem:* Given a curve, find its slope everywhere.
2) *The backward problem:* Given a curve's slope everywhere, find the curve.
3) *The area problem:* Given a curve, find the area under it.

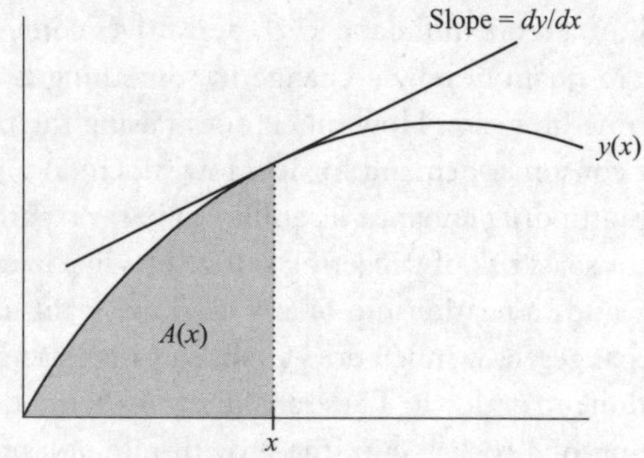

The diagram shows the graph of a generic function $y(x)$. I haven't said what x and y represent because it doesn't matter. The picture is completely general. It shows a curve in the plane. That curve could represent any function of one variable and so could apply to any branch of mathematics or science where such functions arise, which is essentially everywhere. The significance of its slope and area will be explained later. For now, just think of them as what they are: a slope and an area. The kind of thing that geometers would worry about.

We can view this curve in two ways, one old and one new. In the early seventeenth century, before calculus arrived, such curves were viewed as geometrical objects. They were considered fascinating in their own right. Mathematicians wanted to quantify their geometrical properties. Given a curve, they wanted to be able to figure out the slope of its tangent line at each point, the arc length of the curve, the area beneath the curve, and so on. In the twenty-first century, we are more interested in the function that produced the curve, which models some natural phenomenon or technological process that manifested itself in the curve. The curve is data, but something deeper underlies it. Today we think of the curve as footprints in the sand, a clue to the process that made it. That process—modeled by a function—is what we are interested in, not the traces it left behind.

This collision between these two points of view is how the mystery of curves collided with the mysteries of motion and change. It's how ancient geometry collided with modern science. Even though we are in modern times now, I've chosen to draw the picture from the older perspective because the *xy* plane is so familiar. It offers the clearest way to grasp the three central problems of calculus, because all three can be readily visualized when we pose them in geometric terms. (The same ideas can also be reformulated in terms of motion and change using dynamical ideas like speed and distance instead of curves and slope, but we will do that later, once we have a better grasp on the geometry.)

The questions should be interpreted in the sense of functions. In other words, when I speak about the slope of the curve, I don't just mean at one specific point. I mean at an arbitrary point x. The slope changes as we move along the curve. Our goal is to understand how it changes as a function of x. Similarly, the area under the curve depends on x. I've shown it shaded in gray, and labeled it with the symbol $A(x)$. That area should also be regarded as a function of x. As we increase x, the vertical dashed line slides to the right, and the area expands. So the area depends on which x we choose.

These, then, are the three central problems. How can we figure out the changing slope of a curve? How can we reconstruct a curve from its slope? And how can we figure out the changing area beneath the curve?

As phrased in the context of geometry, those questions might sound pretty dry. But once we reinterpret them in the real world, from the twenty-first-century point of view as problems about motion and change, they become phenomenally wide-reaching. Slopes measure rates of change; areas measure the accumulation of change. As such, slopes and areas arise in every field—physics, engineering, finance, medicine, you name it, anywhere that change is an abiding concern. Understanding the problems and their solutions opens up the universe of modern quantitative thinking, at least about functions of one variable. For the sake of full disclosure, I should mention there's much more to calculus than that; there are functions of many variables, differential equations, and the like. All in good time. We'll get to those later.

This chapter is concerned with functions of one variable and their derivatives (their rates of change), starting with functions that change at a constant rate and then moving on to the knottier issue of functions that change at a changing rate. That's where differential calculus really shines—in making sense of ever-changing change.

Once we've gotten comfortable with rates of change, we'll be ready to tackle the accumulation of change, the more challenging topic of the next chapter. There it will be revealed that the forward problem and the backward problem, as different as they seem, are twins separated at birth, a shocker called the fundamental theorem of calculus. It revealed that rates of change and the accumulation of change are much more closely related than anyone had suspected, a discovery that unified the two halves of calculus.

But first, let's begin at the beginning with rates.

Linear Functions and Their Constant Rates

Many situations in everyday life are described by linear relationships in which one variable is proportional to another. For example:

1) Last summer my older daughter, Leah, got her first job, at a clothing store in the mall. She earned $10 an hour, so when she worked for two hours, she made $20. More generally, when she worked for t hours, she made y dollars, where $y = 10t$.
2) A car drives down the highway at 60 miles per hour. Thus, after one hour it goes 60 miles. After two hours, it goes 120 miles. After t hours, it goes $60t$ miles. The relationship here is $y = 60t$, where y is the number of miles driven in t hours.
3) According to the Americans with Disabilities Act, a wheelchair-accessible ramp must not rise by more than 1 inch for every 12 inches of horizontal run. For a ramp with this maximum permissible gradient, the relationship between rise and run is $y = x/12$, where y is the rise and x is the run.

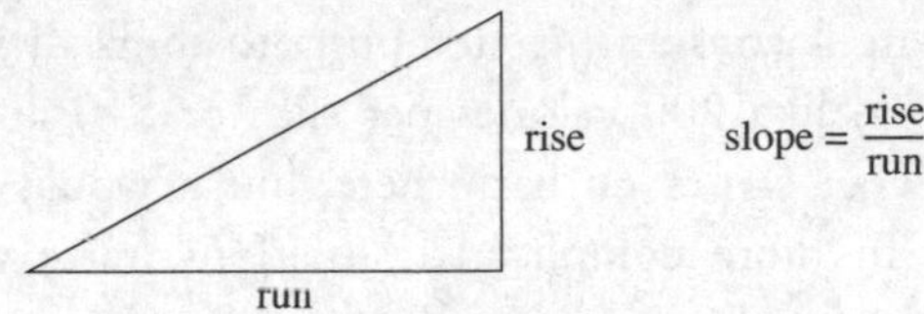

In each of these linear relationships, the dependent variable changes at a constant rate with respect to the independent variable. My daughter's rate of pay was a constant $10 per hour. The car's speed is a constant 60 miles per hour. And the wheelchair-accessible ramp has a constant slope, defined as its rise over run, equal to 1⁄12. The same is true of that cinnamon-raisin bread I like to eat; it delivers calories at a constant rate of 200 calories per slice.

In the technical jargon of calculus, a *rate* always means a quotient of two changes: a change in y divided by a change in x, written in symbols as $\Delta y/\Delta x$. For example, if I eat two more slices of bread, I pack on another 400 calories. Thus the corresponding rate is

$$\frac{\Delta y}{\Delta x} = \frac{400 \text{ calories}}{2 \text{ slices}}$$

which simplifies to 200 calories per slice. No surprise there. But what's interesting to observe is that this rate is constant. It's the same no matter how many slices I've already eaten.

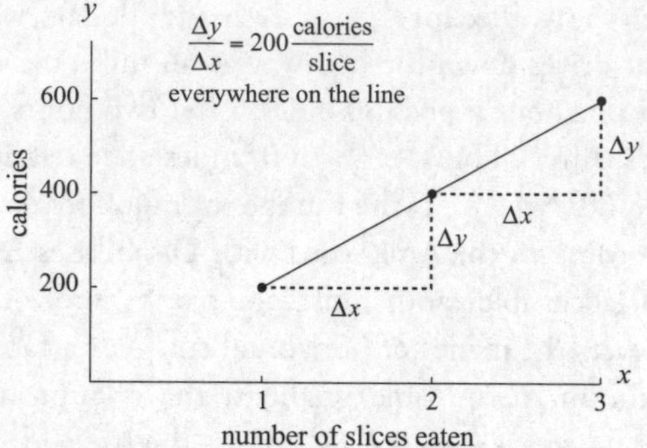

When a rate is constant, it's tempting to think of it as simply being a *number*, like 200 calories per slice or $10 an hour or a slope of 1⁄12. That causes no harm here, but it would get us into trouble later. In more complicated situations, rates will not be constant. For example, consider a walk through a rolling landscape, where some parts of the hike are steep and others are flat. On a rolling landscape, slope is a function of position. It would be a mistake to think of it as a mere number. Likewise, when a car accelerates or when a planet orbits the sun, its speed changes incessantly. Then it's vital to regard speed as a function of time. So we should get in that habit now. We should stop thinking of rates of change as numbers. *Rates are functions.*

The potential confusion arises because the rate functions are constant for the linear relationships we've been considering. That's why it does no harm to treat them as numbers in a linear context. They don't change as we change the independent variable. My daughter's rate of pay is $10 an hour, no matter how much she works, and the slope of the ramp is 1⁄12 everywhere along its length. But don't let that fool you. Those rates are still functions. They just happen to be constant functions. The graph of a constant function is a flat line, as shown here for the cinnamon-raisin bread with its constant payload of 200 calories per slice.

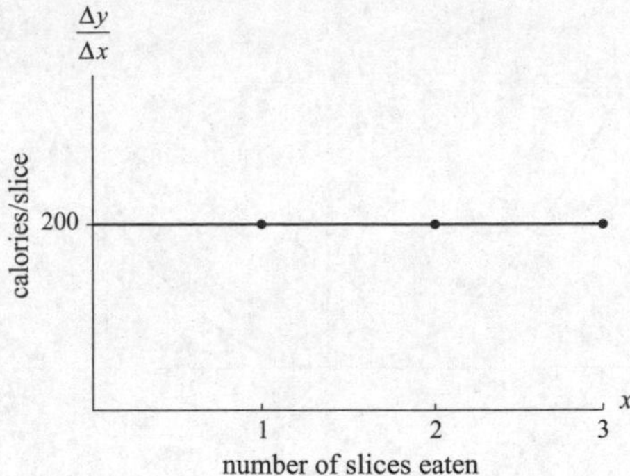

When we deal in the next section with a relationship that is not linear, we will see that it generates a curve, not a line, when graphed in the *xy* plane. Either way, a line or a curve always reveals a lot about the relationship that produced it. It's like a relationship's mug shot or signature. It's a clue that reveals what made it.

Notice the distinction between a function and the graph of the function. A function is a disembodied rule that eats *x*s and spits out *y*s and does so uniquely, one y for each x. In that sense, a function is incorporeal. There's nothing to look at when you look at a function. It's a ghostly entity, an abstract rule. For example, the rule might be "Feed me a number and I will return 10 times the number." By contrast, the graph of a function is a visible, almost tangible thing, a shape you can see. Specifically, the graph of the function I just described would be a line through the origin with a slope of 10, defined by the equation $y = 10x$. But the function itself is not the line. The function is the rule that produces the line. To make a function manifest itself, you need to feed it an x, let it spit out a y, and repeat that for all *x*s and plot the results. When you do that, the function itself stays invisible. What you're seeing is its graph.

A Nonlinear Function and Its Changing Rate

When a function is not linear, its rate of change $\Delta y/\Delta x$ is not constant. In geometrical terms, that means the graph of the function is a curve with a slope that changes from point to point. As an example, consider the parabola shown below.

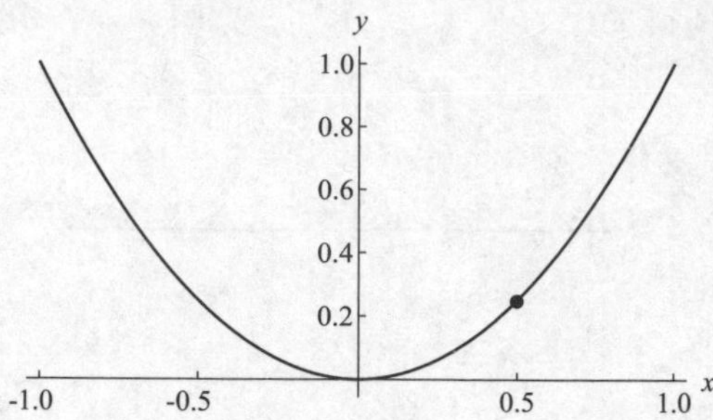

It's the curve $y = x^2$, which corresponds to the simplest nonlinear button on the calculator, the squaring function x^2. This example will give us some practice with the definition of a derivative as the slope of the tangent line and also clarify why limits enter that definition.

Inspecting the parabola, we see that some parts of it are steep and some parts are relatively flat. The flattest part of all occurs at the bottom of the parabola at the point where $x = 0$. There we can see, without doing any work, that the derivative must be zero. It has to be zero because the tangent line at the bottom is evidently the x-axis. Viewed as a ramp, that line is no rise and all run and hence has a slope of zero.

But at other points on the parabola, it's not immediately obvious what the slope of the tangent line should be. In fact, it's not obvious at all. To figure it out, let's do an Einstein-style thought experiment. We'll imagine what we would see if we could zoom in on an arbitrary point (x, y) on the parabola as if we were making photographic enlargements of that point, always keeping it in the center of our field of view. It's like we're looking at a piece of the curve under a microscope and increasing the magnification progressively. As we zoom in closer and closer, that piece of the parabola should begin to look straighter and straighter. In the limit of *infinite* magnification (which amounts to zooming in on an *infinitesimal* piece of the curve around the point of interest), that magnified piece should approach a straight line. If it does, that limiting straight line is defined as the *tangent line* at that point on the curve, and its slope is defined as the *derivative* there.

Notice that we are using the Infinity Principle here—we are trying to make a complicated curve simpler by chopping it into infinitesimal straight pieces. This is what we always do in calculus. Curved shapes are hard; straight shapes are easy, even if there are infinitely many of them and even if they are infinitesimally small. Calculating a derivative in this way is a quintessential calculus move and one of the most fundamental applications of the Infinity Principle.

To conduct the thought experiment, we need to select a point on the curve to zoom in on. Any point will do, but a numerically convenient choice is the point that lies on the parabola above $x = ½$. In the diagram above, I've marked that point with a dot. In the *xy* plane it lies at

$$(x, y) = \left(\tfrac{1}{2}, \tfrac{1}{4}\right)$$

or, in decimal notation, $(x, y) = (0.5, 0.25)$. The reason that y equals ¼ at this point is that, in order to qualify as a point on the parabola, the point must obey $y = x^2$, as all points on the parabola do; after all, this is what defines a point as a member of the parabolic curve. Thus, at $x = ½$, the point must have a y-value of

$$y = x^2 = \left(\tfrac{1}{2}\right)^2 = \tfrac{1}{4}.$$

Now we are ready to zoom in on the point of interest. Place the point $(x, y) = (0.5, 0.25)$ at the center of the microscope. With the help of computer graphics, zoom in on a little piece of the curve surrounding that point. The first magnification is shown here.

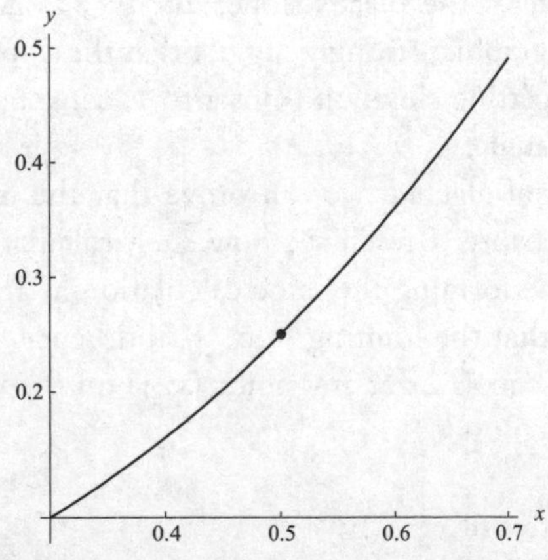

The overall shape of the parabola is lost in this magnified view. Instead, we just see a slightly curved arc. This small piece of the parabola, which lies between $x = 0.3$ and 0.7, appears a lot less curved than the parabola as a whole.

Zoom in further by blowing up the piece between $x = 0.49$ and 0.51. This new enlargement looks even straighter than the last one did, though it's not truly straight, since it's still a portion of the parabola.

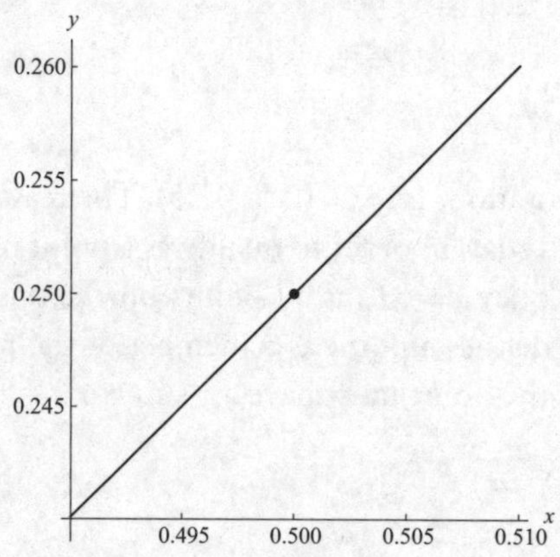

The trend is clear. As we keep zooming in, the pieces look straighter. By measuring the rise over run, $\Delta y / \Delta x$, for this almost-straight piece and zooming in closer and closer, we are effectively taking the limit of the piece's slope, $\Delta y / \Delta x$, as Δx goes to zero. The computer graphics strongly suggest that the slope of the almost-straight line is getting closer and closer to 1, corresponding to a line at a 45-degree angle.

With a bit of algebra, we can prove that the limiting slope is *exactly* 1. (In chapter 8 we'll see how such calculations are done.) Furthermore, performing the same calculation at any x, not just at $x = \frac{1}{2}$, reveals that the limiting slope—and hence the slope of the tangent line—equals $2x$ at any point (x, y) on the parabola. Or in the lingo of calculus:

The derivative of x^2 is $2x$.

Tempting as it is to prove this derivative rule before moving on, for now let's accept it and see what it means. For one thing, it says that at the dot where $x = ½$, the slope should equal $2x = 2 \times (½) = 1$, which is just what we saw in the computer graphics. It also predicts that at the bottom of the parabola at $x = 0$, the slope should be 2×0, which is zero, and we've already seen that's correct too. Finally, the $2x$ formula predicts that the slope should increase as we ascend the parabola to the right; when x gets bigger, the slope ($= 2x$) should also get bigger, which means the parabola should get steeper, and it does.

Our experiment with the parabola helps us understand a couple of caveats about derivatives. A derivative is defined only if a curve approaches a limiting straight line as we zoom in on it. That won't be the case for certain pathological curves. For example, if a curve has a V shape with a sharp corner at one point, then when we zoom in on that point, it will continue to look like a corner. The corner never goes away, no matter how much we magnify the curve. It will never look straight there. Because of this, a V-shaped curve does not have a well-defined tangent line or a slope at the corner, and hence it does not have a derivative there.

However, when a curve *does* look increasingly straight when we zoom in on it sufficiently at any point, that curve is said to be *smooth*. Throughout this book, I have been assuming that the curves and processes of calculus are smooth, just as the early pioneers did. In modern calculus, however, we have learned how to cope with curves that are not smooth. The inconveniences and pathologies of non-smooth curves sometimes arise in applications due to sudden jumps or other discontinuities in the behavior of a physical system. For example, when we flip a switch in an electrical circuit, the current goes from not flowing at all to suddenly flowing significantly. A graph of current versus time would show an abrupt, almost-vertical rise approximated by a discontinuous jump as the current turns on. Sometimes it's more convenient to model that abrupt transition as a truly discontinuous jump, in which case the current as a function of time will not have a derivative at the moment the switch flipped.

Much of the first course in calculus in high school or college is devoted to calculating derivative rules like the one above for x^2 but for the other buttons on the calculator, like "the derivative of sin x equals cos x" or "the derivative of ln x equals $1/x$." For our purposes, however, it's more important to understand the idea of the derivative and to see how its abstract definition applies in practice. For that, let's turn to the real world.

Derivatives as Rates of Change of Day Length

In chapter 4, we looked at data on seasonal changes of day length. Although our purpose at the time was to illustrate ideas about sine waves, curve fitting, and data compression, we can now repurpose those data to illuminate variable rates of change and bring derivatives down to earth in another setting.

The earlier data concerned the number of minutes of daylight—the time between sunrise and sunset—in New York City on each day of the year in 2018. The relevant derivative in this context is the rate at which the days lengthened or shortened from one day to the next. On January 1, for instance, the time from sunrise to sunset was 9 hours, 19 minutes, and 23 seconds. On January 2 it got a little longer: 9 hours, 20 minutes, and 5 seconds. That extra 42 seconds of daylight (equivalent to 0.7 minutes) was a measure of how rapidly the days were lengthening on that particular day of the year. They were getting longer at a rate of about 0.7 minutes per day.

For comparison, consider the rate of change two weeks later, on January 15. Between that day and the next, the amount of daylight increased by 90 seconds, corresponding to a rate of lengthening of 1.5 minutes per day, more than twice the rate of 0.7 measured two weeks earlier. Thus, the days were not only lengthening in January; they were lengthening *faster* with each passing day.

This welcome trend continued for the next several weeks. Daytime kept getting longer—and did so *more rapidly*—with the coming of spring. On the spring equinox, March 20, the rate of increase topped out at a glorious 2.72 minutes of extra sunlight each day. You can spot that day on the earlier graph in chapter 4. It's day 79, about a quarter of the way in from the left, where the wave of day length rises most steeply. That makes sense—where the graph is steepest, it's climbing most rapidly, which means the derivative is largest and the days are lengthening as quickly as possible. All of this happens on the first day of spring.

For a melancholy contrast, consider the shortest days of the year. They pack a double whammy. In those dark days of winter, the days are not only depressingly short; they also do not change much from one day to the next, which only adds to the torpor. But this also makes sense. The shortest days occur at the bottom of the wave of day length, and at the bottom, the wave is flat (otherwise it wouldn't be a bottom; it would be improving or worsening). But because it is flat at the bottom, its derivative is zero there, which means its rate of change grinds to a halt, at least momentarily. On days like that, it can feel like spring will never come.

I've highlighted two times of year that have emotional meaning for many of us, around the spring equinox and the winter solstice, but it's even more instructive to consider the year as a whole. To track the seasonal variations in the rate of change of day length, I've computed it at periodic intervals throughout the year, starting on January 1 and continuing every two weeks after that. The results are plotted in the graph below.

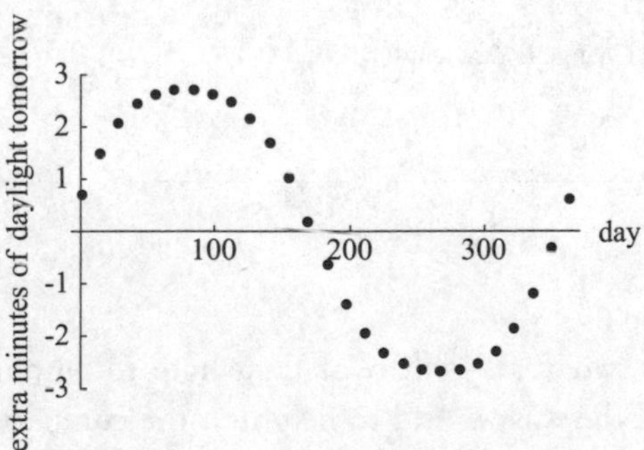

The vertical axis shows the daily rate of change, that is, the additional minutes of daylight from one day to the next. The horizontal axis shows what day it is, with days numbered from 1 (January 1) to 365 (December 31).

The rate of change bobs up and down like a wave. It starts out positive in the late winter and early spring, when the days are getting longer, and peaks around day 79 (the spring equinox, March 20). As we already know, that's when the days are lengthening most rapidly, around 2.72 minutes per day. But after that, it's all downhill. The rate starts to drop and goes negative after the summer solstice on day 172 (June 21). It becomes negative because the days start shortening then; the next day has fewer hours of daylight than the current one. The rate bottoms out around September 22 when the light is fading fastest, and it stays negative (but not as negative) until the winter solstice on day 355 (December 21) when the days start getting longer again, even if imperceptibly.

It's fascinating to compare this wave to the wave we met earlier in chapter 4. When they're plotted together and rescaled to have comparable amplitudes, here's what they look like.

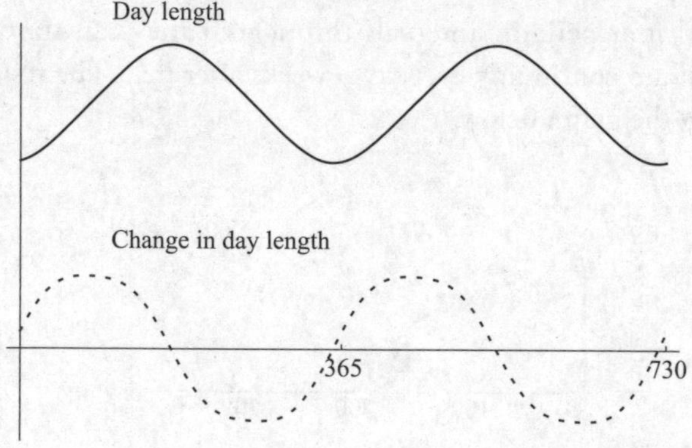

(I'm showing two years' worth of data here to emphasize the repetitiveness of the waves. And to heighten the comparison between them, I've also connected the dots and removed the numbers from the vertical axis to focus more attention on the waves' shape and timing.)

The first thing to notice is that the waves are out of sync. They don't peak simultaneously. The wave of day length peaks around halfway through the year, whereas its rate of change peaks about three months earlier. That amounts to a quarter of a cycle earlier, given that each wave takes twelve months to complete its up-and-down movement.

The other thing to notice is that the waves resemble each other, with slight differences. Although they show clear family ties, the dashed wave is less symmetrical than the solid one and its peaks and troughs are flatter.

I'm going into all this because these real-world waves offer a glimpse, as through a glass darkly, of a remarkable property of sine waves, namely, when a variable follows a perfect sine-wave pattern, its rate of change is also a perfect sine wave timed a quarter of a cycle ahead. This self-regeneration property is unique to sine waves. No other kinds of waves have it. It could even be taken as a *definition* of sine waves. In that sense, our data hint at a marvelous phenomenon of rebirth inherent in perfect sine waves. (We will have more to say about this when it comes up again in connection with Fourier analysis, a powerful offshoot of calculus that has led to some of its most exciting applications today.)

Let me try to give you some insight into where the quarter-cycle shift comes from. The same concept explains why sine waves beget sine waves when we compute their rates of change. The key is that sine waves are connected to uniform circular motion. Recall that when a point moves around a circle at a constant speed, its up-and-down motion traces a sine wave in time. (For that matter, so does its left and right motion.) With that in mind, consider the diagram below.

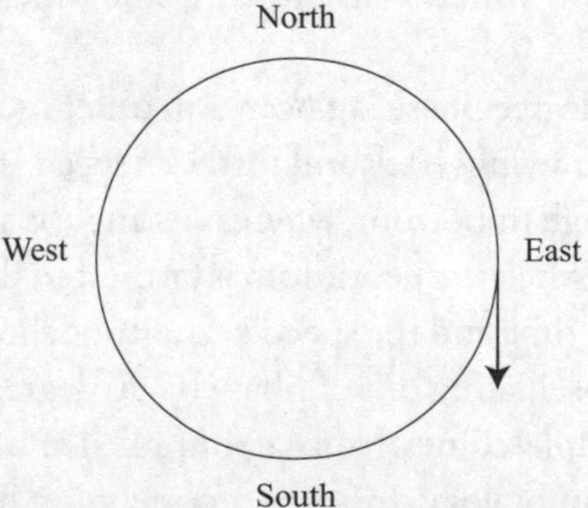

It shows a point moving clockwise around a circle. The point is not supposed to represent anything physical or astronomical. It's not the Earth orbiting the sun, and it doesn't have anything to do with the seasons. It's just an abstract point moving around a circle. Its eastward displacement (or "eastiness," for short) increases and decreases like a sine wave. When the point reaches its maximum eastiness, as shown in the diagram, that's analogous to the maximum of a sine wave, or the longest day of the year. The question is: When the point is maximally east and the sine wave is at its peak eastiness, what happens next? As the diagram shows, at its easternmost point, the point heads *south*, as indicated by the downward arrow. But south is 90 degrees away from east on a compass, and 90 degrees is a quarter of a cycle. Aha! That's where the quarter-cycle offset comes from. Because of the geometry of a circle, there's always a quarter-cycle offset between any sine wave and the wave derived from it as its derivative, its rate of change. In this analogy, the point's direction of travel is like its rate of change. It determines where the point will go next and hence how it changes its location. Moreover, this compass heading of the arrow itself rotates in a circular fashion at a constant speed as the point goes around the circle, so the compass heading of the arrow follows a sine-wave pattern in time. And since the compass heading is like the rate of change, voilà! The rate of change follows a sine-wave pattern too. That's the self-regeneration property we were trying to understand. Sine waves beget sine waves with a 90-degree shift. (Experts will realize that I'm trying to explain without formulas why the derivative of the sine function is the cosine function, which is itself just a sine function shifted by a quarter cycle.)

A similar 90-degree phase lag occurs in other oscillating systems. When a pendulum swings back and forth, its speed is at its maximum when it goes through its bottom, whereas its angle is at its maximum a quarter cycle later when the pendulum is farthest to the right. A graph of the angle versus time and the speed versus time shows two approximate sine waves, oscillating out of phase by 90 degrees.

Another example comes from a simplified model of predator-prey interactions in biology. Imagine a population of sharks preying on a population of fish. When the fish are at their maximum population level, the shark population grows at its maximum rate because there are so many fish to eat. The shark population continues to climb and reaches its own maximum level a quarter cycle later, by which time the fish population has dropped, having been preyed on so severely a quarter cycle earlier. An analysis of this model shows that the two populations oscillate out of phase by 90 degrees. Similar predator-prey oscillations are seen elsewhere in nature, for example, in annual fluctuations of Canadian hare and lynx populations as recorded by fur-trapping companies in the 1800s (though the real explanation for those oscillations is undoubtedly more complicated, as is often the case in biology).

Returning to the day-length data, we see that, alas, they are not perfect sine waves. They're also an inherently discrete set of points, just one per day, with no data existing in between. As such, they do not provide the sort of continuum of points that calculus insists on. So for our final example of a derivative, let's turn to a case where we can collect data with as much resolution as we like, right down to the millisecond.

Derivatives as Instantaneous Speeds

The evening of August 16, 2008, was windless in Beijing. At ten thirty, the eight fastest men in the world lined up for the Olympic finals of the 100-meter dash. One of them, a twenty-one-year-old Jamaican sprinter named Usain Bolt, was a relative newcomer to this event. Known more as a 200-meter man, he'd begged his coach for years to let him try running the shorter race, and over the past year he'd become very good at it.

He didn't look like the other sprinters. He was gangly, 6 feet, 5 inches (1.96 meters) tall, with a long, loping stride. As a boy he had focused on soccer and cricket until his cricket coach noticed his speed and suggested that he try out for track. As a teenager he kept improving as a runner, but he never took the sport or himself too seriously. He was goofy and mischievous and had a fondness for practical jokes.

On that night in Beijing, after all the athletes had been introduced and finished mugging for the TV cameras, the stadium got quiet. The sprinters placed their feet in the blocks and crouched into position. An official called out, "On your marks. Set," and then fired the starting pistol.

Bolt shot out of the blocks, but not quite as explosively as the other Olympians. His slower reaction time left him seventh out of eight near the start. Gaining speed, by thirty meters he moved up to the middle of the pack. Then, still accelerating like a bullet train, he put daylight between himself and the rest of the field.

At eighty meters, he glanced to his right to see where his main competitors were. When he realized how far ahead he was, he slowed down visibly, dropped his arms to his sides, and slapped his chest as he cruised across the finish line. Some commentators saw this as bragging, others as gleeful celebration, but in any case, Bolt clearly didn't feel the need to run hard at the end, which led to speculation about just how fast he could have run. As it was, even with his celebration (and an untied shoelace) he set a new world record of 9.69 seconds. One official criticized him for being unsportsmanlike, but Bolt didn't mean any disrespect. As he later told reporters, "That's just me. I like to have fun, just stay relaxed."

How fast did he run? Well, 100 meters in 9.69 seconds translates to $100/9.69 = 10.32$ meters per second. In more familiar units, that's about 37 kilometers per hour, or 23 miles per hour. But that was his *average* speed over the whole race. He went slower than that at the beginning and end and faster than that in the middle.

More detailed information is available from his split times recorded every 10 meters down the track. He covered the first 10 meters in 1.83 seconds, corresponding to an average speed of 5.46 meters per second there. His fastest splits occurred at 50 to 60 meters, 60 to 70 meters, and 70 to 80 meters. He blazed through those 10-meter sections in 0.82 seconds each, for an average speed of 12.2 meters per second. In the final 10 meters, when he eased up and broke form, he decelerated to an average speed of 11.1 meters per second.

Human beings have evolved to spot patterns, so instead of poring over numbers like we've just been doing, it's usually more informative to visualize them. The following graph shows the elapsed times at which Bolt crossed 10 meters, 20 meters, 30 meters, and so on, up to the 9.69 seconds it took him to cross the finish line at the 100-meter mark.

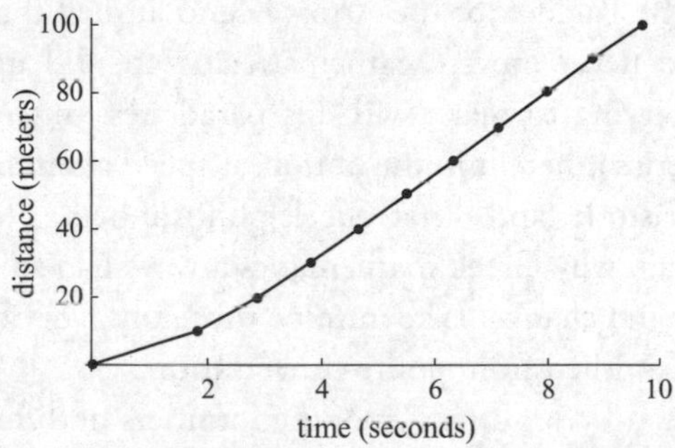

I've connected the dots with straight lines as a guide to the eye, but keep in mind that only the dots are real data. Together the dots and the line segments between them form a polygonal curve. The slopes of the segments are shallowest on the left, corresponding to Bolt's lower speed at the start of the race. They bend upward as they move to the right; that means he's accelerating. Then they join to form a nearly straight line, indicating the high and steady speed that he maintained for most of the race.

It's natural to wonder at what time he was running his

absolute fastest and where on the track that occurred. We know that his fastest *average* speed, over a 10-meter section, occurred somewhere between 50 and 80 meters, but an average speed over 10 meters is not quite what we want; we are interested in his *peak* speed. Imagine that Usain Bolt was wearing a speedometer. At what precise moment was he running the fastest? And exactly how fast was that?

What we're looking for here is a way of measuring his instantaneous speed. The concept seems almost paradoxical. At any instant, Usain Bolt was at precisely one place. He was frozen, as in a snapshot. So what would it mean to speak of his speed at that instant? Speed can only occur over a time interval, not in a single instant.

The enigma of instantaneous speed goes far back in the history of mathematics and philosophy, to around 450 BCE with Zeno and his redoubtable paradoxes. Recall that in his paradox of Achilles and the tortoise, Zeno claimed that a faster runner could never overtake a slower runner, despite what Usain Bolt proved that night in Beijing. And in the Paradox of the Arrow, Zeno argued that an arrow in flight could never move. Mathematicians are still unsure what point he was trying to make with his paradoxes, but my guess is that the subtleties inherent in the notion of speed at an instant troubled Zeno, Aristotle, and other Greek philosophers. Their uneasiness may explain why Greek mathematics always had so little to say about motion and change. Like infinity, those unsavory topics seem to have been banished from polite conversation.

Two thousand years after Zeno, the founders of differential calculus solved the riddle of instantaneous speed. Their intuitive solution was to define instantaneous speed as a limit—specifically, the limit of average speeds taken over shorter and shorter time intervals.

It's like what we did when we zoomed in on the parabola. There, we approximated a smaller and smaller piece of a smooth curve with a straight line. Then we asked what happens in the limit of infinite magnification. By studying the limiting value of the line's slope, we were able to define the derivative at a particular point on the smoothly curving parabola.

Here, by analogy, we would like to approximate something changing smoothly in time: Usain Bolt's distance down the track. The idea is to replace the graph of his distance versus time with a polygonal curve changing at a constant average speed over short time intervals. If the average speed on each interval approaches a limit as those time intervals get shorter and shorter, that limiting value is what we mean by the instantaneous speed at a given time. Like slope at a point, speed at an instant is a derivative.

For all this to succeed, we have to assume his distance down the track varied smoothly. Otherwise the limit we're investigating won't exist, and neither will the derivative. The results won't approach anything sensible as the intervals get shorter. But did his distance actually vary smoothly as a function of time? We don't know for sure. The only data we have are discrete samples of Bolt's elapsed times at each of the ten-meter markers on the track. To estimate his instantaneous speed, we need to go beyond the data and make an educated guess about where he was at times in between those points.

A systematic way to make such a guess is known as interpolation. The idea is to draw a smooth curve between the data available. In other words, we want to connect the dots, not by straight-line segments as we've already done, but by the most plausible smooth curve that goes through the dots, or at least that goes very close to them. The constraints we impose on this curve are that it should be taut and not undulate too much; it should pass as close to all the dots as possible; and it should show that Bolt's initial speed was zero, since we know he was motionless when he was in the crouch position. There are many different curves that meet these criteria. Statisticians have devised a host of techniques for fitting smooth curves to data. All of them give similar results, and since they all involve a bit of guesswork anyway, let's not bother too much about which one to use.

Here's one example of a smooth curve that meets all the requirements.

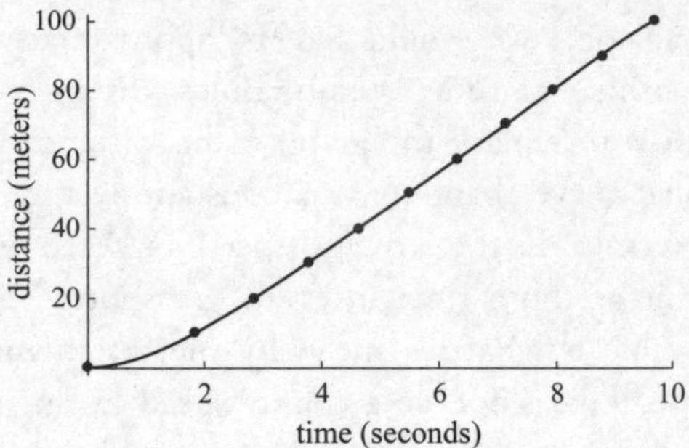

Since the curve is smooth by design, we can calculate its derivative at every point. The resulting graph gives us an estimate of Usain Bolt's velocity at each instant of his record-setting race that night in Beijing.

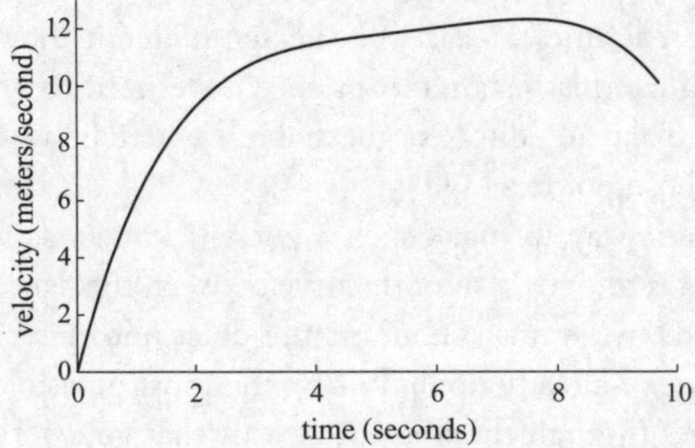

It indicates that Bolt reached a top speed of around 12.3 meters per second at about the three-quarter point in the race. Until then, he'd been accelerating, gaining speed at each moment. After that he decelerated, so much so that his speed dropped to 10.1 meters per second as he crossed the finish line. The graph confirms what everyone saw; Bolt slowed down dramatically near the end, especially in the last twenty meters, when he relaxed and celebrated.

The next year, at the 2009 World Championships in Berlin, Bolt put an end to the speculation about how fast he could go. No chest thumps this time. He ran hard to the finish and shattered his Beijing world record of 9.69 seconds with an even more astonishing time of 9.58 seconds. Because of the great anticipation surrounding this event, biomechanical researchers were on hand with laser guns, similar to the radar guns used by police to catch speeders. These high-tech instruments allowed the researchers to measure the sprinters' positions a hundred times a second. When they computed Bolt's instantaneous speed, this is what they found:

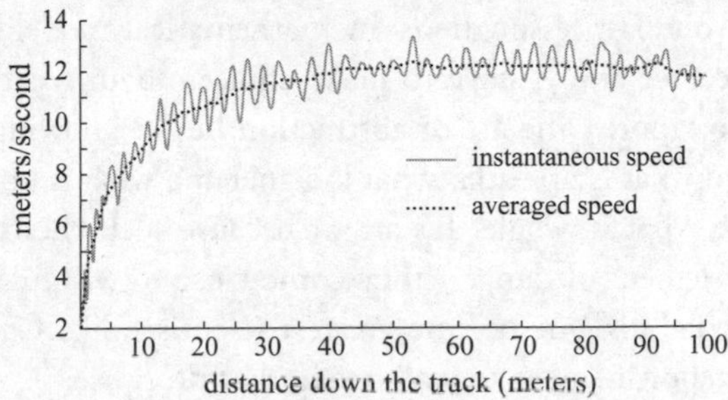

The little wiggles on the overall trend represent the ups and downs in speed that inevitably occur during strides. Running, after all, is a series of leapings and landings. Bolt's speed changed a little whenever he landed a foot on the ground and momentarily braked, then propelled himself forward and launched himself airborne again.

Intriguing as they are, these little wiggles are annoying and bothersome to a data analyst. What we really wanted to see was the trend, not the wiggles, and for that purpose, the earlier approach of fitting a smooth curve to the data was just as good and arguably better. After collecting all that high-resolution data and observing the wiggles, the researchers had to clean them off anyway. They filtered them out to unmask the more meaningful trend.

To me, these wiggles hold a larger lesson. I see them as a metaphor, a kind of instructional fable about the nature of modeling real phenomena with calculus. If we try to push the resolution of our measurements too far, if we look at any phenomenon in excruciatingly fine detail in time or space, we will start to see a breakdown of smoothness. In Usain Bolt's speed data, the wiggles took a smooth trend and made it look as bushy as a pipe cleaner. The same thing would happen with any form of motion if we could measure it at the molecular scale. Down at that level, motion becomes jittery and far from smooth. Calculus would no longer have much to tell us, at least not directly. Yet if what we care about are the overall trends, smoothing out the jitters may be good enough. The enormous insight that calculus has given us into the nature of motion and change in this universe is a testament to the power of smoothness, approximate though it may be.

There's one last lesson here. In mathematical modeling, as in all of science, we always have to make choices about what to stress and what to ignore. The art of abstraction lies in knowing what is essential and what is minutia, what is signal and what is noise, what is trend and what is wiggle. It's an art because such choices always involve an element of danger; they come close to wishful thinking and intellectual dishonesty. The greatest scientists, like Galileo and Kepler, somehow manage to walk along that precipice.

"Art," said Picasso, "is a lie that makes us realize truth." The same could be said for calculus as a model of nature. In the first half of the seventeenth century, calculus began to be used as a powerful abstraction of motion and change. In the second half of that century, the same kinds of artistic choices—the lies that revealed the truth—prepared the way for a revolution.

7

The Secret Fountain

In the second half of the seventeenth century, Isaac Newton in England and Gottfried Wilhelm Leibniz in Germany changed the course of mathematics forever. They took a loose patchwork of ideas about motion and curves and turned it into a calculus.

Notice the indefinite article. When Leibniz introduced the word *calculus* in this context in 1673, he spoke of "a calculus" and sometimes, more affectionately, "my calculus." He was using the word in its generic sense, a system of rules and algorithms for performing computations. Later, after his system was brought to a high polish, its accompanying article was upgraded to the definite, and the field became known as *the* calculus. But now, sad to say, its articles and possessives have all gone away. What remains is calculus, humdrum and gray.

Articles aside, the word *calculus* itself has stories to tell. It comes from the Latin root *calx*, meaning a small stone, a reminder of a time long ago when people used pebbles for counting and thus for calculations. The same root gives us words like *calcium, chalk,* and *caulk.* Your dentist might use the word *calculus* to refer to that gunk on your teeth, the tiny pebbles of solidified plaque the hygienist scrapes off when you go for a cleaning. Doctors use the same word for gallstones, kidney stones, and bladder stones. In a cruel irony, both Newton and Leibniz, the pioneers of calculus, died in excruciating pain while suffering from calculi—a bladder stone for Newton, a kidney stone for Leibniz.

Areas, Integrals, and the Fundamental Theorem

Although calculus had once been about counting with stones, by the time of Newton and Leibniz it was devoted to curves and their newfangled analysis through algebra. Thirty years earlier, Fermat and Descartes had discovered how to use algebra to find the maxima, minima, and tangents of curves. What remained elusive were the *areas* of curves or, more precisely, the areas of regions bounded by curves.

This *area problem*, classically known as the quadrature, or squaring, of curves, had consumed and frustrated mathematicians for two thousand years. Many ingenious tricks had been devised to solve particular cases, from Archimedes's work on the area of the circle and the quadrature of the parabola to Fermat's solution for the area under the curve $y = x^n$. But what was lacking was a system. Area problems were tackled on an ad hoc basis, case by case, as if the mathematician were starting over each time.

The same difficulty beset problems about the volumes of curved solids and the lengths of curved arcs. Indeed, Descartes thought arc lengths were beyond human comprehension. In his book on geometry, he wrote, "The ratio which exists between straight and curved lines is not known, and even cannot, in my judgment, be known by man." All these problems—areas, arc lengths, and volumes—required infinite sums of infinitesimally small pieces. In modern parlance, they all involved *integrals*. Nobody had a surefire system for any of them.

This is what changed after Newton and Leibniz. They independently discovered and proved a fundamental theorem that made such problems routine. The theorem connected areas to slopes and thereby linked integrals to derivatives. It was astonishing. Like a twist out of a Dickens novel, two seemingly distant characters were the closest of kin. Integrals and derivatives were related by blood.

The impact of this fundamental theorem was breathtaking. Almost overnight, areas became tractable. Questions that earlier savants had strained to solve could now be dispatched in a matter of minutes. As Newton wrote to a friend of his, "There is no curved line expressed by any equation . . . but I can in less than half a quarter of an hour tell whether it may be squared." Realizing how incredible this claim would sound to his contemporaries, he continued, "This may seem a bold assertion . . . but it's plain to me by the fountain I draw it from, though I will not undertake to prove it to others."

Newton's secret fountain was the fundamental theorem of calculus. Although he and Leibniz weren't the first to notice this theorem, they get the credit for it because they were the first to prove it in general, recognize its overwhelming utility and importance, and build an algorithmic system around it. The methods they developed are now commonplace. Integrals have been defanged and turned into homework exercises for teenagers.

Right now, millions of students in high school and college all around the world are grinding away on their calculus problem sets, solving integral after integral with the help of the fundamental theorem. Yet many of them are oblivious to the gift they've been given. Perhaps understandably so—it's like the old joke about the fish who asks his friend, "Aren't you grateful for water?" to which the other fish says, "What's water?" Students in calculus are swimming in the fundamental theorem all the time, so naturally they take it for granted.

Visualizing the Fundamental Theorem with Motion

The fundamental theorem can be understood intuitively by thinking about the distance traveled by a moving body like a runner or a car. By acquainting ourselves with this way of thinking, we'll learn what the fundamental theorem says, why it's true, and why it matters. It's not just a trick for finding areas. It's the key to predicting the future of anything we care about (in the cases where we can) and for unlocking the secrets of motion and change in the universe.

The fundamental theorem occurred to Newton when he looked at the area problem dynamically. His brainstorm was to invite time and motion into the picture. Let the area flow, said he. Let it expand continuously.

The simplest illustration of his idea takes us back to the familiar problem of a car moving at a constant speed for which distance equals rate times time. As elementary as this example may be, it still captures the essence of the fundamental theorem, so it's a good place to start.

Imagine a car cruising down the highway at 60 miles per hour. If we plot its distance versus time and, beneath that, its speed versus time, the resulting distance and speed graphs look like this:

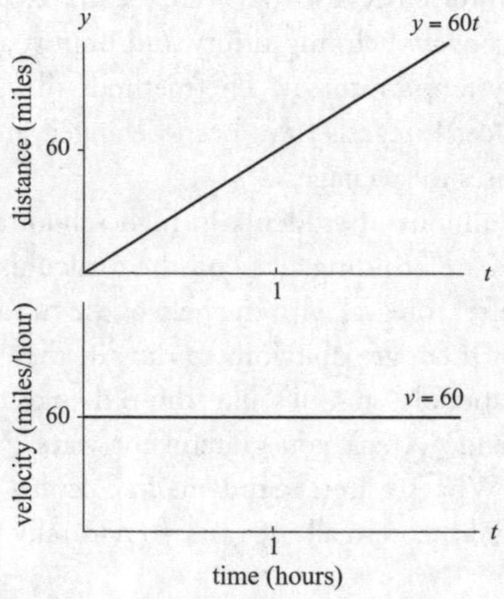

Look at distance versus time first. After one hour the car has traveled 60 miles, and after two hours, 120 miles, and so on. In general, distance and time are related by $y = 60t$, where y denotes the distance the car has traveled up to time t. I'll refer to $y(t) = 60t$ as the *distance function*. As shown in the top diagram, the graph of the distance function is a straight line with a slope of 60 miles per hour. That slope tells us the car's speed at every instant if we didn't already know it. In a harder problem, the speed might fluctuate, but here it's a simple constant function, $v(t) = 60$ at all t, graphed as the flat line in the bottom panel of the diagram. (Here v stands for velocity, a synonym for speed.)

Having seen how speed manifests itself on the distance graph (as the slope of the line), we now turn the question around and ask: How does distance reveal itself on the speed graph? In other words, is there some visual or geometric feature of the speed graph that would allow us to infer how far the car has traveled up to any given time *t*? Yes. *The distance traveled is the area accumulated under the speed curve (the flat line) up to time t.*

To see why, suppose the car drives for some particular amount of time, say a half an hour. In that case the distance traveled would be 30 miles, since distance equals rate times time and 60 × ½ = 30. The cool thing, and the point of all this, is that we can read off that distance as the area of the gray rectangle under the flat line between times $t = 0$ and $t = ½$ hour.

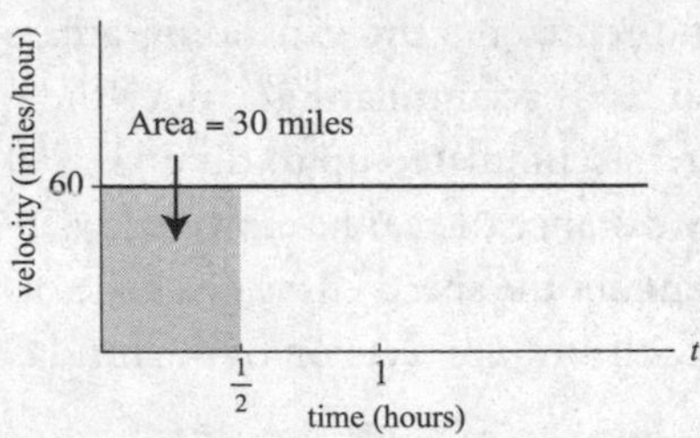

The rectangle's height of 60 miles per hour times its base of ½ hour gives the rectangle's area, 30 miles, which recovers the distance traveled, as claimed.

The same reasoning works for *any* time *t*. The base of the rectangle then becomes *t* and its height is still 60 so its area is $60t$, and, indeed, that's the distance we were expecting to find, $y = 60t$.

So, at least in this example, where speed was perfectly constant and the speed curve was simply a flat line, the key to recovering distance from speed was to compute the area under the speed curve. Newton's insight was this equality between area and distance *always* holds, even if the speed is not constant. *No matter how erratically something moves, the area accumulated under its speed curve up to time t always equals the total distance it has traveled up to that time.* That's one version of the fundamental theorem. It seems too easy to be true, but it *is* true.

Newton was led to it by thinking of area as a flowing, moving quantity, not as a frozen measure of a shape, as was then customary in geometry. He brought time into geometry and viewed it like physics. If he were alive today, perhaps he would have visualized the picture above as an animation, more like a flipbook than a snapshot. To do this, look at the picture above one last time, but now imagine it as a single frame in a movie or a single page in a flipbook. As the animation plays in our minds, what would we see the gray rectangle do? We would see it expanding sideways. Why? Because its base has length t, which grows as time passes. If we could make a frame for each time and replay them in sequence, like flipping the pages of a flipbook, the animated version of the gray rectangle would look like it was stretching to the right. It would resemble a piston expanding or a syringe lying on its side, pulling gray fluid into itself.

That gray fluid represents the expanding area of the rectangle. We think of the area as "accumulating" under the speed curve $v(t)$. In this case, the area accumulated up to time t is $A(t) = 60t$, and that coincides with the distance the car has traveled, $y(t) = 60t$. Thus, the accumulated area under the speed curve gives the distance as a function of time. That's the motion version of the fundamental theorem.

Constant Acceleration

We're working our way up to Newton's general geometric version of the fundamental theorem, which is phrased in terms of an abstract curve $y(x)$ and the area $A(x)$ accumulated beneath it. The idea of area accumulation is the key to explaining the theorem, but I realize this idea takes some getting used to, so let's apply it to one more concrete problem about motion before tackling the abstract geometric case.

Consider an object that moves with a constant acceleration. That means it keeps going faster and faster with a speed that ramps up at a constant rate. It's roughly like what would happen if you were to floor the gas pedal in your car, starting from rest. After one second, the car might be going, perhaps, 10 miles per hour; after two seconds, 20 miles per hour; after three seconds, 30 miles per hour, and so on. In this hypothetical example, the car always gains

10 miles per hour with each passing second. This rate of change of speed, 10 miles per hour per second, is defined as the car's *acceleration*. (For simplicity, we are ignoring the fact that a real car has a top speed it can't exceed and that its acceleration might not be strictly constant when you floor the gas pedal.)

In our idealized example, the car's speed at each moment is given by the linear function $v(t) = 10t$. Here the number 10 signifies the car's acceleration. If the acceleration were some other constant, say a, the formula would generalize to

$$v(t) = at.$$

What we want to know is, for a car peeling out like this, how far does it go between time 0 and time t? In other words, how does its distance from the starting point increase as a function of time? It would be a horrible blunder to invoke the middle-school formula of distance equals rate times time because that formula is valid only when the rate—the car's speed—is constant, which it certainly isn't here. On the contrary, in this problem the car's speed is ramping up at every instant. We are no longer in the sleepy world of constant speed. This is the thrilling world of constant acceleration.

Scholars in the Middle Ages already knew the answer. William Heytesbury, a philosopher and logician at Merton College, Oxford, solved the problem around 1335, and Nicole Oresme, a French cleric and mathematician, elucidated it further and analyzed it pictorially around 1350. Unfortunately their works were not widely studied and were soon forgotten. About two hundred and fifty years later, Galileo demonstrated experimentally that constant acceleration is not a purely academic assumption. It is actually how heavy objects like iron balls move when they fall freely near the surface of the Earth or when they roll down a gently sloping ramp. In both cases, a ball's speed v really does grow in proportion to time, $v = at$, as expected for motion with a constant acceleration.

Next, knowing that the speed grows linearly according to $v = at$, how does the *distance* grow? The fundamental theorem says the distance traveled equals the area accumulated under the speed curve up to time t. And since here the speed curve is the sloping line $v = at$, the relevant area is easy to compute. It's given by the area of the triangle below.

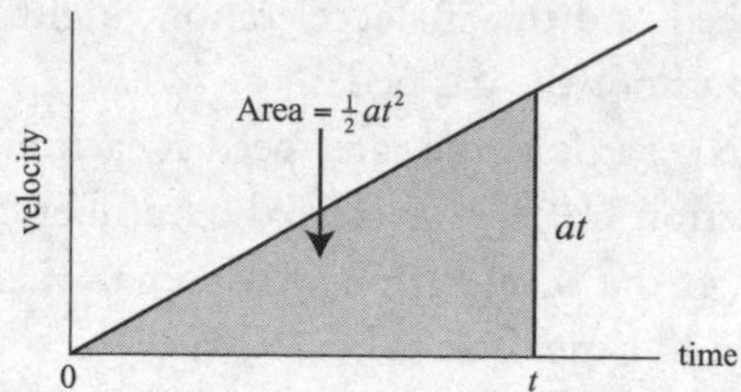

Like the gray rectangle in the previous problem, the gray triangle here is expanding as time passes. The difference is that the rectangle expanded only horizontally whereas this triangle is expanding in both directions. To compute how fast its area is expanding, observe that at any time t the triangle's base is t and its height is the body's current speed, $v = at$. Since the area of a triangle is half its base times its height, the accumulated area equals $\frac{1}{2} \times t \times at = (\frac{1}{2})at^2$. By the fundamental theorem, that area under the speed curve tells us how far the body has traveled:

$$y(t) = \frac{1}{2}at^2 .$$

Hence, for a body that starts from rest and accelerates uniformly, the distance traveled increases in proportion to the *square* of the time elapsed. This is exactly what Galileo discovered experimentally and expressed in such a charming fashion with his law of odd numbers, as we saw in chapter 3. The scholars in the Middle Ages knew it too.

But what was not known in the Middle Ages, or even in the time of Galileo, was how the velocity would behave when the acceleration was *not* simply constant. In other words, given a body moving with an arbitrary acceleration $a(t)$, what could one say about its speed $v(t)$?

This is like the backward problem I mentioned in the last chapter. It's a tricky question. To understand it properly, it's crucial to appreciate what we know and don't know.

The acceleration is defined as the rate of change of speed. So if we were given the speed $v(t)$, finding the corresponding acceleration

$a(t)$ would be easy. That's called *solving the forward problem*. We could solve it by computing the rate of change of the given speed function in much the same way that we calculated the slope of the parabola in the last chapter by placing it under the microscope. Finding a rate of change of a known function requires nothing more than invoking the definition of the derivative and applying the many rules for calculating derivatives of various functions.

But what makes the backward problem so tricky is that we are *not* given the speed function. On the contrary, we are being asked to *find* the speed function. We are assuming that we have been given its rate of change—its acceleration—as a function of time, and we are trying to figure out what speed function has that acceleration function as its given rate of change. How can we go *backward* to infer an unknown speed from its known rate of change? It's like a children's game: "I'm thinking of a speed function whose rate of change is such and such. What speed function am I thinking of?"

The same puzzle of having to reason backward arises when we try to infer distance from speed. Just as acceleration is the rate of change of speed, speed is the rate of change of distance. Reasoning forward is easy; if we know a moving body's distance as a function of time, as we did in the case of Usain Bolt running down the track in Beijing, it's easy to calculate the body's speed at every instant. We did that calculation in the last chapter. But reasoning backward is difficult. If I told you how fast Usain Bolt was running at every instant in the race, could you infer where he was on the track at each moment? More generally, given an arbitrary speed function $v(t)$, could you infer the corresponding distance function $y(t)$?

Newton's fundamental theorem shed light on this very difficult backward problem of inferring an unknown function from its given rate of change and in many cases solved it completely. The key was to reframe it as a question about areas that flow and expand.

A Paint-Roller Proof of the Fundamental Theorem

The fundamental theorem of calculus was the culmination of eighteen centuries of mathematical thought. By dynamic means, it answered a static geometric question that Archimedes could have asked in ancient Greece in 250 BCE or that could have occurred to Liu Hui in China in 250 CE or to Ibn al-Haytham in Cairo in 1000 or to Kepler in Prague in 1600.

Consider a shape like the gray region shown here.

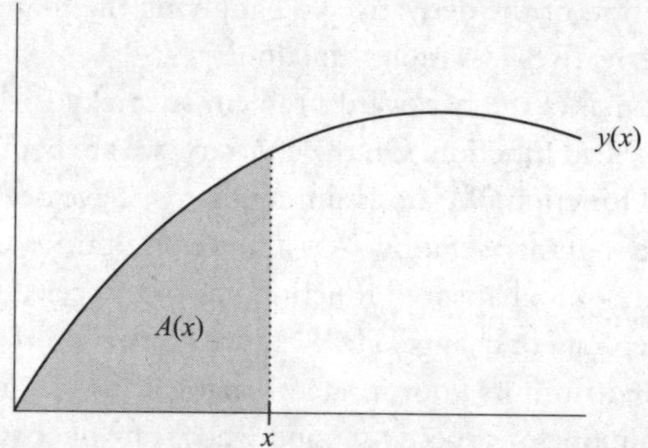

Is there a way to compute the exact area of an arbitrary shape like this, given that the curve on top could be almost anything? In particular, it needn't be a classic curve. It could be an exotic new curve defined by an equation in the *xy* plane, the jungle opened up by Fermat and Descartes. Or what if the curve was defined by something of physical interest, like a trajectory of a moving particle or the path of a light ray—was there any way to find the area under such an arbitrary curve and do it systematically? This was the *area problem*, the third central problem of calculus I mentioned earlier and the most pressing mathematical challenge of the mid-1600s. It was the last remaining puzzle in the mystery of curves. Isaac Newton approached it from a new direction, using ideas inspired by the mysteries of motion and change.

Historically, the only way to solve problems like this had been to be clever. You had to find some cunning way to slice a curved region into strips or smash it into shards and then reassemble the pieces in your mind or weigh them on an imaginary seesaw, as Archimedes had done. But around 1665 Newton gave the area problem its first major advance in nearly two millennia. He incorporated the insights of Islamic algebra and French analytic geometry but went far beyond them.

The first step, according to his new system, was to lay the area down in the *xy* plane and determine an equation for its curved top. This required computing how far the curve was above the *x*-axis, one vertical slice at a time (as indicated by the dotted vertical line in the diagram) to obtain the corresponding *y*. That computation converted the curve into an equation relating *y* to *x*, which made it susceptible to the instruments of algebra. Thirty years earlier, Fermat and Descartes had already understood this much and had used these techniques to find tangent lines to curves, a huge breakthrough in itself.

But what they missed was that tangent lines per se were not that important. More important than such lines were their *slopes*, for it was slopes that led to the concept of the derivative. As we saw in the last chapter, the derivative arose very naturally in geometry as the slope of a curve. And derivatives also arose in physics as other rates of change, such as speeds. Thus, derivatives suggested a link between slopes and speeds and, more broadly, between geometry and motion. Once the idea of the derivative was firmly in Newton's mind, its power to bridge geometry and motion made the final breakthrough possible. It was the derivative that finally unlocked the area problem.

The deeply hidden connections among all these ideas—slopes and areas, curves and functions, rates and derivatives—emerged from the shadows when Newton looked at the area problem dynamically. In the spirit of our earlier work in the last two sections, ponder the diagram above and imagine sliding *x* to the right at a constant speed. You could even think of *x* as time; Newton often did. Then the area of the gray region changes continuously as *x* moves. Because that area depends on *x*, it should be regarded as a function of *x*, so we write it as $A(x)$. When we want to stress that this area is a function of *x* (as opposed to a frozen number), we refer to it as the *area accumulation function*, or sometimes just the *area function*.

My calculus teacher in high school, Mr. Joffray, had a memorable metaphor for this fluid scenario, with its sliding x and its changing area. He asked us to imagine a magical paint roller moving sideways. As it rolls steadily to the right, it paints the region under the curve gray.

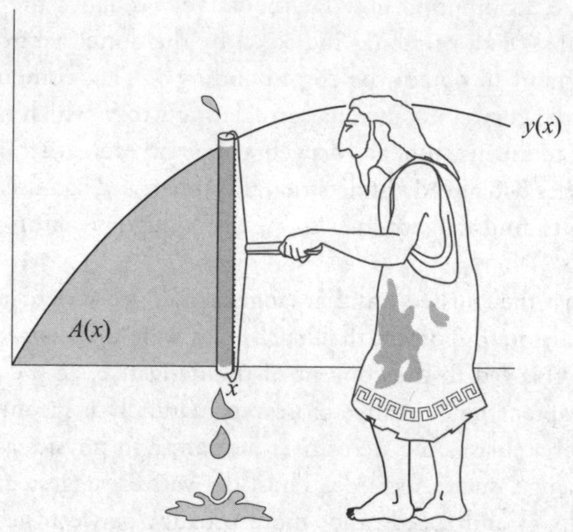

The dotted line at x marks the current position of this imaginary roller as it rolls to the right. Meanwhile, to ensure that the region is painted neatly, the roller instantly and magically shrinks or stretches in the vertical direction, exactly as needed to reach the curve on top and the x-axis on the bottom without ever crossing those boundaries. The magical aspect is that it always adjusts its length to $y(x)$ as it rolls, so it paints the area immaculately.

Having set up this far-fetched scenario, we ask: At what *rate* does the gray area expand as x moves to the right? Or, equivalently, what's the rate at which paint is being laid down when the roller is at x? To answer that, think about what happens in the next infinitesimal interval of time. The roller rolls to the right through some infinitesimal distance dx. Meanwhile, as it traverses that tiny distance, it keeps its length y in the vertical direction almost perfectly constant, since there's almost no time for it to change its length during the infinitesimally brief roll (a fine point that we'll discuss in the next chapter). During that brief interval, it paints what is essentially a tall, thin rectangle of height y, infinitesimal width dx, and infinitesimal area $dA = y\,dx$. Dividing this equation by dx then reveals the rate at which area accumulates. It is given by

$$\frac{dA}{dx} = y\,.$$

This tidy formula says that the total painted area under the curve increases at a rate given by the current length y of the paint roller. It makes sense; the longer the roller currently is, the more paint it lays down in the next instant, and so the faster the area accumulates.

With a little more effort we could show that this geometric version of the theorem is equivalent to the motion version we used earlier, which stated that the area accumulated under a speed curve equals the distance traveled by a moving body. But we have more urgent tasks ahead. We need to understand what the theorem means, why it matters, and how it ultimately changed the world.

The Meaning of the Fundamental Theorem

The diagram below summarizes what we've just learned.

$$A(x) \xrightarrow{\text{derivative}} y(x) \xrightarrow{\text{derivative}} \frac{dy}{dx}$$

area under curve	curve	slope of curve

It shows the three functions we're interested in and the relationships between them. The given curve is in the middle, its unknown slope is on the right, and its unknown area is on the left. As we saw in chapter 6, these are the functions that occur in the three central problems of calculus. Given the curve y, we are trying to figure out its slope and its area.

I hope the diagram now makes clear why I referred to finding the slope as "the forward problem." To find the slope from the curve, we simply follow the arrow on the right by moving forward along it. We compute the derivative of y to find its slope. That's the straightforward problem (1) we discussed in the last chapter.

What we did not know before and what we have just learned from the fundamental theorem is that the area A and the curve y are *also* related by a derivative—the fundamental theorem has revealed that the derivative of A is y. This is a stupendous fact. It gives us an avenue for figuring out the area underneath an arbitrary curve, the age-old mystery that stumped the greatest minds for almost two thousand years. The picture now suggests a path to the answer. But before we uncork the champagne, we should realize that the fundamental theorem does not quite give us what we want. It does not give us the area directly. But it tells us how to obtain it.

The Holy Grail of Integral Calculus

As I've tried to make clear, the fundamental theorem doesn't fully solve the area problem. It provides information about the rate at which the area changes, but we still need to infer the area itself.

In terms of symbols, the fundamental theorem tells us that $dA/dx = y$, where $y(x)$ is our given function. We're still left with the chore of finding an $A(x)$ that satisfies this equation. Wait a minute—this means we're suddenly faced with the *backward problem* again! It's a remarkable turn of events. We were trying to solve the area problem, central problem number 3 on our list in chapter 6, and suddenly we're being confronted by the backward problem, central problem number 2 on the list. I'm calling it the backward problem because, as the diagram above shows, finding A from y means going upstream against the arrow, going *backward* against the derivative. In this setting the children's game might go something like this: "I'm thinking of an area function $A(x)$ whose derivative is $12x + x^{10} - \sin x$. What function am I thinking of?"

Developing methods to solve the backward problem, not just for $12x + x^{10} - \sin x$ but for any curve $y(x)$, became the holy grail of calculus. More precisely, it became the holy grail of *integral* calculus. Solving the backward problem would allow the area problem to be solved once and for all. Given any curve $y(x)$, we'd know the area $A(x)$ underneath it. By solving the backward problem, we'd also solve the area problem. This is what I meant about those two problems being separated-at-birth twins and two sides of the same coin.

A solution to the backward problem would also have much larger implications, for the following reason: An area is, from an Archimedean standpoint, an infinite sum of infinitesimal rectangular strips. As such, an area is an *integral*. It's the integrated collection of all the pieces put back together, an accumulation of infinitesimal change. And just as derivatives are more important than slopes, integrals are more important than areas. Areas are crucial to geometry; integrals are crucial to *everything*, as we'll see in the chapters ahead.

One way to approach the difficult backward problem is to ignore it. Shunt it aside. Replace it with the easier forward problem (given A, compute its rate of change dA/dx; by the fundamental theorem, we know that this must equal the y we're seeking). This forward problem is much easier because we know where to start. We can start with a known area function $A(x)$ and then crank out its rate of change by applying standard formulas for derivatives. The resulting rate of change dA/dx then must play the role of the partner function y; that's what the fundamental theorem assures us: $dA/dx = y$. Having done all that, we now have a pair of partner functions, $A(x)$ and $y(x)$, which represent an area function and its associated curve. The hope is that if we are lucky enough to stumble across a problem where we need to find the area under this particular curve $y(x)$, its corresponding area function will be its partner $A(x)$. It's not a systematic approach and it works only if we happen to get lucky, but at least it's a start and it's easy. To increase our odds of success, we can make a big lookup table that lists hundreds of area functions and their associated curves as $\left(A(x), y(x)\right)$ pairs. Then the sheer size and diversity of that table will improve our chances of stumbling across the pair we need to solve a genuine area problem of interest. Having found the necessary pair, we wouldn't need to do any further work. The answer would be right there in the table.

For example, in the next chapter we'll see that the derivative of x^3 is $3x^2$. We'll obtain that result by solving a forward problem, simply taking a derivative. What's wonderful about it, however, is that it tells us that x^3 could play the role of $A(x)$, and $3x^2$ could play the role of $y(x)$. Without breaking a sweat, we've solved the area problem for $3x^2$ (should we ever happen to be interested in it). Continuing in this fashion, we can fill in the table with other power functions of x. Similar calculations would show that the derivative of x^4 is $4x^3$, the derivative of x^5 is $5x^4$, and in general the derivative of x^n is nx^{n-1}. These are all easy solutions of the forward problem for power functions. Thus the columns of the table would look like this:

Curve $y(x)$	Its area function $A(x)$
$3x^2$	x^3
$4x^3$	x^4
$5x^4$	x^5
$6x^5$	x^6
$7x^6$	x^7

In his college notebook, a twenty-two-year-old Isaac Newton wrote out similar tables for himself.

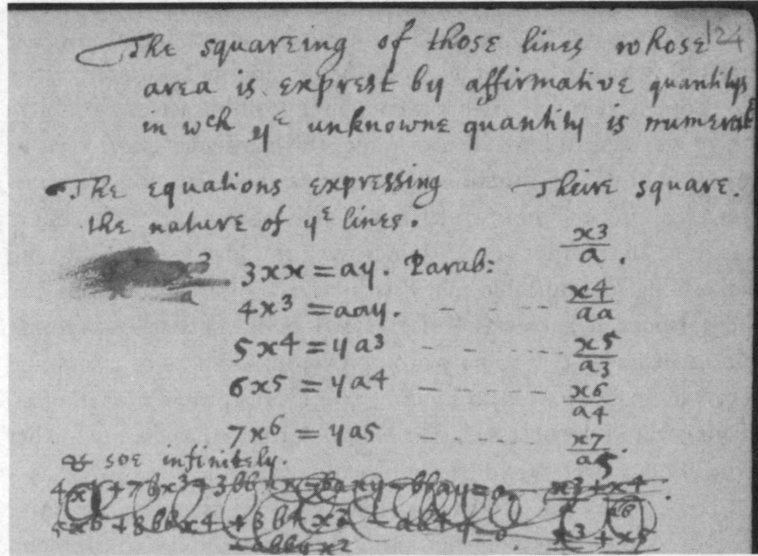

The squaring of those lines whose 24
area is exprest by affirmative quantitys
in wch ye unknowne quantity is numerat

The equations expressing the nature of ye lines.	Their square.
3xx = ay. Parab:	x3/a.
4x3 = aay.	x4/aa
5x4 = ya3	x5/a3
6x5 = ya4	x6/a4
7x6 = ya5	x7/a5.

& soe infinitely.

Notice that his language is a bit different from ours. The curves in the left column are "The equations expressing the nature of ye lines." Their area functions are "Theire square" (because he views the area problem as the "squaring of curves"). He also feels the need to insert various powers of *a*, an arbitrary unit of length, to ensure that all quantities have the proper number of dimensions. For example, his bottom right $A(x)$, five lines down from the top of the list, is x^7/a^5 (instead of our simpler x^7) because in his mind, it represents an area and hence needs to have units of length squared. All of this comes a few pages after "A method whereby to square those crooked lines which may be squared"—the birth announcement of the fundamental theorem of calculus. Armed with that theorem, Newton filled many more pages with lists of "crooked lines" and their "squares." In Newton's hands, the machinery of calculus was beginning to whir.

The next task, a fantasy, really, was to find a method to square *any* curve, not just power functions. Perhaps it doesn't sound like a particularly scintillating fantasy. But that's because it's so general. Let me put it this way: This problem contains the distilled essence of what makes integral calculus so challenging. If this problem could be solved, it would be like setting off a chain reaction. It would be like toppling dominoes; one problem after another would fall. If this problem could be solved, it could be used to answer the question that Descartes thought was beyond human comprehension, finding the arc length of an arbitrary curve. It would be possible to find the area of any amoeba-shaped region in the plane. It would be possible to calculate the surface areas, volumes, and centers of gravity of spheres, paraboloids, urns, barrels, and all other surfaces made by spinning a curve around an axis, like a vase on a potter's wheel. The classic problems about curved shapes that Archimedes pondered and that another eighteen centuries of mathematical talent pondered after him would all become tractable instantly, in a single stroke.

Not only that, but certain problems of prediction would be overcome as well. Predicting the position of a moving object far into the future—for instance, where a planet will be at a certain point in its orbit, even a planet that obeys a different force of attraction than the one operating in our own universe—would become possible if just this one problem could be solved. That's what I mean by calling it the holy grail of integral calculus. Many, many other problems boil down to solving this one. If it goes, they all go.

This is why it was so important to be able to find the area under an arbitrary curve. Because of its intimate connection to the backward problem, the area problem is not just about area. It's not just about shape or the relationship between distance and speed or anything that narrow. It's completely general. From a modern perspective, the area problem is about predicting the relationship between anything that changes at a changing rate and how much that thing builds up over time. It's about the fluctuating inflow to a bank account and the accumulated balance of money in it. It's about the growth rate of the world's population and the net number of people on Earth. It's about the changing concentration of a chemotherapy drug in a patient's blood and the accumulated exposure to that drug over time. That total exposure affects how potent the chemo will be, as well as how toxic. Area matters because the future matters.

Newton's new mathematics was exquisitely suited to a world in flux. Accordingly, he christened it *fluxions*. He spoke of fluent quantities (which we now think of as functions of time) and their fluxions (their derivatives, their rates of change in time). He identified two central problems:

1) Given the fluents, how can one find their fluxions? (This is equivalent to the forward problem we mentioned earlier, the easy problem of finding the slope of a given curve or, more generally, finding the rate of change or derivative of a known function, the process known today as *differentiation.*)
2) Given the fluxions, how can one find their fluents? (This is equivalent to the backward problem and the key to the area problem; it is the difficult problem of inferring a curve from its slope or, more generally, inferring an unknown function from its rate of change, the process known today as *integration.*)

Problem 2 is much harder than problem 1. It's also much more important for prediction and for tapping into the code of the universe. Before we look at how far Newton got on it, let me try to clarify why it's so hard.

Local Versus Global

The reason why integration is so much harder than differentiation has to do with the distinction between local and global. Local problems are easy. Global problems are hard.

Differentiation is a local operation. As we've seen, when we are calculating a derivative, it's like we're looking under a microscope. We zoom in on a curve or a function, repeatedly magnifying the field of view. As we zoom in on that little local patch, the curve appears to become less and less curved. We see a blown-up version of the curve, a tiny ramp, almost perfectly straight, with a rise Δy and a run Δx. In the limit of infinite magnification, it approaches a certain straight line, the tangent line at the point in the center of the microscope. The slope of that limiting line gives us the derivative there. The role of the microscope is to let us focus on the part of the curve we care about. Everything else gets ignored. That's the sense in which finding the derivative is a local operation. It discards all details outside the infinitesimal neighborhood of a point, the only point of interest.

Integration is a global operation. Instead of a microscope, we are now using a telescope. We are trying to peer far off into the distance —or far ahead into the future, although in that case we need a crystal ball. Naturally, these problems are a lot harder. All the intervening events matter and cannot be discarded. Or so it would seem.

Let me offer an analogy to bring out these distinctions between local and global, between differentiation and integration, and to clarify why integration is so hard and so scientifically important. The analogy takes us back to Beijing and Usain Bolt's record-breaking race. Recall that to find his speed at each instant, we fit a smooth curve to the data showing his position on the track as a function of time. Then, to find his speed at a certain point, say 7.2 seconds into the race, we used the fitted curve to estimate his position a short time later, say at 7.25 seconds, and then looked at the change in distance divided by the change in time to estimate his speed at that moment. These were all local calculations. The only information they used was how he was running in the few hundredths of a second around that given time. Everything he did in the rest of the race, before and after, was irrelevant. That's what I mean by local.

By contrast, think about what would be involved if we were handed an infinitely long spreadsheet showing his speed at every moment in the race and asked to reconstruct where he was 7.2 seconds after the start. As he comes out of the blocks, we could use his initial speed to estimate where he was at, let's say, a hundredth of a second later by using distance equals rate times time to advance him down the track. From that new position and that new elapsed time, we could again advance him down the track over the next hundredth of a second with the corresponding speed and the corresponding distance he would cover. On and on, inching down the track, accumulating information one hundredth of a second at a time, we could update his position throughout the race. It would be an arduous grind. Computationally, I mean. This is what makes a global calculation so difficult. We need to compute every step to get to a desired answer far into the future, in this case 7.2 seconds after the starting gun went off.

But imagine if we could somehow fast-forward and zap straight to the instant we cared about—now, *that* would be useful. And that is exactly what a solution to the backward problem of integration would achieve. It would give us a shortcut, a wormhole through time. It would convert a global problem into a local one. That's why solving the backward problem would be like finding the holy grail of calculus.

It was first solved, as so many things are, by a student.

A Lonesome Boy

Isaac Newton was born in a stone farmhouse on Christmas Day 1642. Apart from the date, there was nothing auspicious about his arrival. He was born premature and was so tiny, it was said, he could fit inside a quart mug. He was also fatherless. The elder Isaac Newton, a yeoman farmer, had died three months earlier, leaving behind barley, furniture, and some sheep.

When little Isaac was three, his mother, Hannah, remarried and left him in the care of his maternal grandparents. (His mother's new husband, Reverend Barnabas Smith, insisted on this arrangement; he was a wealthy man twice her age and wanted a young wife but not a young son.) Understandably, Isaac resented his stepfather and felt abandoned by his mother. Later in life, on a list of sins he'd committed before the age of nineteen, he included this entry: "13. Threatning my father and mother Smith to burne them and the house over them." The next entry was darker: "14. Wishing death and hoping it to some." And then this: "15. Striking many. 16. Having uncleane thoughts words and actions and dreamese."

He was a troubled, lonely little boy with no companions and too much time on his hands. He pursued scholarly investigations on his own, building sundials in the farmhouse, measuring the play of light and shadows on the wall. When he was ten, his mother returned, widowed again, with three new children in tow, two daughters and a son. She sent Isaac away to a school in Grantham, eight miles up the road, too far for him to walk each day. He boarded with Mr. William Clark, an apothecary and chemist, from whom he learned cures and remedies, boiling and mixing, and how to grind with a mortar and pestle. The schoolmaster, Mr. Henry Stokes, taught him Latin, a bit of theology, Greek, Hebrew, and some practical math for farmers about surveying and measuring acreage, as well as some deeper things, like how Archimedes had estimated pi. Although his school reports described him as an idle and inattentive student, when Isaac was alone in his room at night, he drew shapes on the wall, Archimedean diagrams of circles and polygons.

When he was sixteen his mother pulled him out of school and forced him to run the family farm. He hated farming. He allowed his swine to trespass on his neighbors' fields and let his fences fall apart, and he was duly fined by the manor court. He got in fights with his mother and half sisters. He would often lie in the fields and read by himself. He built waterwheels in the stream and studied the whorls they made in the flow.

Finally, his mother did the right thing. At the urging of her brother and schoolmaster Stokes, she allowed Isaac to go back to school. He performed well enough academically that in 1661, he was able to enter Trinity College, Cambridge, as a sizar. Being a sizar meant he had to earn his keep by waiting on tables and serving the richer students. Sometimes he ate their leftovers. (His mother could have afforded to support him, but she didn't.) He made few friends in college, a pattern that would continue for the rest of his life. He never married and, as far as we know, never had a romantic relationship. He rarely laughed.

His first two years of college were taken up with Aristotelian scholasticism, still standard at that time. But then his mind began to stir. He became curious about mathematics after reading a book on astrology. He found he couldn't understand it without knowing some trigonometry and that he couldn't understand trigonometry without knowing some geometry, so he took a look at Euclid's *Elements*. At first all the results seemed obvious to him, but he changed his mind when he came to the Pythagorean theorem.

In 1664 he was awarded a scholarship, and he delved into mathematics in earnest. Teaching himself from six standard texts of the era, he got up to speed on the basics of decimal arithmetic, symbolic algebra, Pythagorean triples, permutations, cubic equations, conic sections, and infinitesimals. Two authors especially enthralled him: Descartes, on analytic geometry and tangents, and John Wallis, on infinity and quadrature.

At Play with Power Series

While poring over Wallis's *Arithmetica Infinitorum* in the winter of 1664–65, Newton chanced upon something magical. It was a new way to find the areas under curves, a way that was both easy and systematic.

In essence, he turned the Infinity Principle into an algorithm. The traditional Infinity Principle says that to compute a complicated area, reimagine it as an infinite series of simpler areas. Newton followed that strategy, but he updated it by using symbols, not shapes, as his building blocks. Instead of the usual shards or strips or polygons, he used powers of a symbol x, like x^2 and x^3. Today we call his strategy *the method of power series.*

Newton viewed power series as a natural generalization of infinite decimals. An infinite decimal, after all, is nothing but an infinite series of powers of 10 and 1⁄10. The digits in the number tell us how much of each power of 10 or 1⁄10 to mix in. For example, the number pi = 3.14 . . . corresponds to this particular mix:

$$3.14\ldots = 3\times 10^0 + 1\times\left(\tfrac{1}{10}\right)^1 + 4\times\left(\tfrac{1}{10}\right)^2 + \cdots.$$

Of course, to write any number in this manner, we need to allow ourselves to use *infinitely* many digits, which is what infinite decimals demand and require. By analogy, Newton suspected he could concoct any curve or function out of infinitely many powers of x. The trick was to figure out how much of each power to mix in. In the course of his studies he developed several methods for finding the right mix.

He hit on his method while thinking about the area of a circle. By making this ancient problem more general, he uncovered a structure within it that nobody had ever noticed before. Rather than restricting his attention to a standard shape, like a whole circle or a quarter circle, he looked at the area of an oddly shaped "circular segment" of width x, where x could be any number from 0 to 1 and where 1 was the radius of the circle.

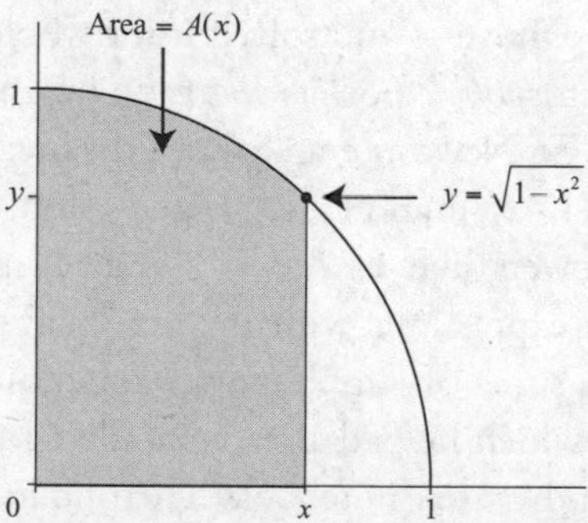

This was his first creative move. The advantage of using the variable x was that it let Newton adjust the shape of the region continuously, as if by turning a knob. A small value of x near 0 would produce a thin, upright segment of the circle, like a thin strip standing on its edge. Increasing x would fatten the segment into a blocky region. Going all the way up to an x-value of 1 would give him the familiar shape of a quarter circle. By dialing x up or down, he could go anywhere he liked in between.

Through a freewheeling process of experimentation, pattern recognition, and inspired guesswork (a style of thinking he learned from Wallis's book), Newton discovered that the area of the circular segment could be expressed by the following power series:

$$A(x) = x - \tfrac{1}{6}x^3 - \tfrac{1}{40}x^5 - \tfrac{1}{112}x^7 - \tfrac{5}{1152}x^9 - \cdots.$$

As for where those peculiar fractions came from or why the powers of x were all odd numbers, well, that was Newton's secret sauce. He cooked it up by an argument that can be summarized as follows. (Feel free to skip the rest of this paragraph if you are not especially interested in his argument. However, if you would like to see the details, check out the notes for references.) Newton began his work on the circular segment by using analytic geometry. He expressed the circle as $x^2 + y^2 = 1$ and then solved for y to get $y = \sqrt{1-x^2}$. Next he argued that the square root was equivalent to a half power and hence that $y = (1 - x^2)^{1/2}$; note the ½ power to the right of the parenthesis. Then, since neither he nor anyone else knew how to find the areas of segments for half powers, he sidestepped the problem—his second creative move—and solved it for whole powers instead. Finding the areas for *whole* powers was easy; he knew how from his reading of Wallis. So Newton cranked out the areas of segments for $y = (1 - x^2)^1$ and $(1 - x^2)^2$ and $(1 - x^2)^3$ and so on, all of which have whole-number powers like 1, 2, and 3 outside their parentheses. He expanded the expressions with the binomial theorem and saw that they became sums of simple power functions, the individual area functions of which he had already tabulated, as we saw on the page from his handwritten notebook. Then he looked for patterns in the areas of the segments as functions of x. Based on what he saw for whole powers, he guessed the answer—his third creative move—for half powers and then checked it in various ways. The answer for the ½ power led him to his formula for $A(x)$, the amazing power series with the peculiar fractions displayed above.

The derivative of the power series for the circular segment then led him to an equally amazing series for the circle itself:

$$y = \sqrt{1-x^2} = 1 - \tfrac{1}{2}x^2 - \tfrac{1}{8}x^4 - \tfrac{1}{16}x^6 - \tfrac{5}{128}x^8 - \cdots.$$

There was much more to come, but already this was remarkable. He'd concocted a circle out of infinitely many simpler pieces—simpler, that is, from the standpoint of integration and differentiation. All its ingredients were power functions of the form x^n, where the power n was a whole number. All the individual power functions had easy derivatives and integrals (area functions). Likewise, the numerical values of x^n could be calculated with simple arithmetic using nothing more than repeated multiplication and could then be combined into a series, again using nothing more than addition, subtraction, multiplication, and division. There were no square roots to take or any other messy functions to worry about. If he could find power series like this for *other* curves besides circles, integrating them would become effortless too.

At barely twenty-two, Isaac Newton had found a path to the holy grail. By converting curves to power series, he could find their areas systematically. The backward problem was a cinch for power functions, given the pairs of functions he had tabulated. So any curve that he could express as a series of power functions was every bit as easy to solve. This was his algorithm. It was tremendously powerful.

Then he tried a different curve, the hyperbola $y = 1/(1+x)$, and found he could write it, too, as a power series:

$$\frac{1}{1+x} = 1 - x + x^2 - x^3 + x^4 - x^5 + \cdots.$$

This series in turn led him to a power series for the area of a segment under the hyperbola from 0 to x, the hyperbolic counterpart of the circular segment he'd studied earlier. It defined a function that he called the hyperbolic logarithm and that today we call the natural logarithm:

$$\ln(1+x) = x - \tfrac{1}{2}x^2 + \tfrac{1}{3}x^3 - \tfrac{1}{4}x^4 + \tfrac{1}{5}x^5 - \tfrac{1}{6}x^6 + \cdots.$$

Logarithms excited Newton for two reasons. First, they could be used to speed up calculations enormously, and second, they were relevant to a controversial problem in music theory he was working on: how to divide an octave into perfectly equal musical steps without sacrificing the most pleasing harmonies of the traditional scale. (In the jargon of music theory, Newton was using logarithms to assess how faithfully an equal-tempered division of the octave could approximate the traditional tuning of just intonation.)

Thanks to the marvels of the internet and the historians at the Newton Project, you can travel back to 1665 right now and watch young Newton at play. (His handwritten college notebook is freely available at http://cudl.lib.cam.ac.uk/view/MS-ADD-04000/.) Look over his shoulder at page 223 (105v in the original) and you'll see him comparing musical and geometrical progressions. Zoom in on the bottom of that page to see how he connects his calculations to logarithms. Then go to page 43 (20r in the original) to watch him "square the hyperbola" and use his power series to calculate the natural logarithm of 1.1 to fifty digits.

What kind of person calculates logarithms by hand to fifty digits? He seemed to be reveling in the newfound strength his power series gave him. When he later reflected on the extravagance of this calculation, he sounded a bit sheepish: "I am ashamed to tell to how many places I carried these computations, having no other business at that time: for then I took really too much delight in these inventions."

If it's any consolation, nobody's perfect. When he first did these computations, Newton made a small arithmetic error. His calculation was correct to only twenty-eight digits. He later caught the error and fixed it.

After his foray with the natural logarithm, Newton extended his power series to the trigonometric functions, which arise whenever circles or cycles or triangles appear, as in astronomy, surveying, and navigation. Here, however, Newton was not the first. More than two centuries earlier, mathematicians in Kerala, India, had discovered power series for the sine, cosine, and arctangent functions. Writing in the early 1500s, Jyesthadeva and Nilakantha Somayaji attributed these formulas to Madhava of Sangamagrama (c. 1350–c. 1425), the founder of the Kerala school of mathematics and astronomy, who derived them and expressed them in verse about two hundred and fifty years before Newton. In a way it makes sense that power series should have been anticipated in India. Decimals were also developed in India, and as we've seen, Newton regarded what he was doing for curves as an analog of what infinite decimals had done for arithmetic.

The point of all this is that Newton's power series gave him a Swiss army knife for calculus. With them, he could do integrals, find roots of algebraic equations, and calculate the values of non-algebraic functions like sines, cosines, and logarithms. As he put it, "By their help analysis reaches, I might almost say, to all problems."

Newton as Mash-Up Artist

I don't believe Newton was consciously aware of it, but in his work on power series he behaved like a mathematical mash-up artist. He approached area problems in geometry via the Infinity Principle of the ancient Greeks and infused it with Indian decimals, Islamic algebra, and French analytic geometry.

Some of his mathematical debts are visible in the architecture of his equations. For example, compare the infinite series of *numbers* that Archimedes used in his quadrature of the parabola,

$$\tfrac{4}{3} = 1 + \tfrac{1}{4} + \tfrac{1}{16} + \tfrac{1}{64} + \cdots,$$

with the infinite series of *symbols* that Newton used in his quadrature of the hyperbola:

$$\frac{1}{1+x} = 1 - x + x^2 - x^3 + x^4 - x^5 + \cdots.$$

If you plug $x = -\tfrac{1}{4}$ into Newton's series, it becomes Archimedes's series. In that sense, Newton's series subsumes Archimedes's as a special case.

What's more, the similarity in their work extends to the geometric problems they considered. Both of them were fond of segments; Archimedes used his number series to square (or find the area of) a parabolic segment, whereas Newton used his jacked-up power series,

$$A_{\text{circular}}(x) = x - \tfrac{1}{6}x^3 - \tfrac{1}{40}x^5 - \tfrac{1}{112}x^7 - \tfrac{5}{1152}x^9 - \cdots,$$

to square a circular segment, and he used a different power series,

$$A_{\text{hyperbolic}}(x) = x - \tfrac{1}{2}x^2 + \tfrac{1}{3}x^3 - \tfrac{1}{4}x^4 + \tfrac{1}{5}x^5 - \tfrac{1}{6}x^6 + \cdots,$$

to square a hyperbolic segment.

Actually, Newton's series were infinitely more powerful than Archimedes's in that they enabled him to find the areas of not just one but a whole continuous infinity of circular and hyperbolic segments. That's what his abstract symbol x did for him. It let him change his problems continuously and effortlessly. It enabled him to tune the shape of segments by sliding x to the left or right so that what appeared to be a single infinite series was in fact an infinite family of infinite series, one for each choice of x. That was the power of power series. They let Newton solve infinitely many problems in a single stroke.

But again, he couldn't have done any of this without standing on the shoulders of giants. He unified, synthesized, and generalized ideas from his great predecessors: He inherited the Infinity Principle from Archimedes. He learned his tangent lines from Fermat. His decimals came from India. His variables came from Arabic algebra. His representation of curves as equations in the xy plane came from his reading of Descartes. His freewheeling shenanigans with infinity, his spirit of experimentation, and his openness to guesswork and induction came from Wallis. He mashed all of this together to create something fresh, something we're still using today to solve calculus problems: the versatile method of power series.

A Private Calculus

While Newton was working on power series during the winter of 1664–65, a terrible pestilence was sweeping north across Europe, moving like a wave, propagating up from the Mediterranean and into Holland. When the bubonic plague reached London, it killed hundreds in a week, and then thousands. In the summer of 1665, Cambridge University temporarily shut down in defense. Newton went home to the family farmhouse in Lincolnshire.

Over the next two years he became the best mathematician in the world. But inventing modern calculus wasn't enough to keep his mind occupied. He also discovered the inverse-square law of gravity and applied it to the moon, invented the reflecting telescope, and showed experimentally that white light is composed of all the colors of the rainbow. He was not yet twenty-five. As he later recalled, "In those days I was in the prime of my age for invention and minded mathematics and philosophy more than at any time since."

In 1667, after the plague abated, Newton returned to Cambridge and continued his solitary studies. By 1671, he had unified the disparate parts of calculus into a seamless whole. He'd developed the method of power series, vastly improved on existing theories of tangent lines by exploiting ideas about motion, found and proved the fundamental theorem, which cracked the area problem, compiled tables of curves and their area functions, and welded all of these into a finely tuned, systematic, computational machine.

But beyond the cloisters of Trinity College, he was invisible. That was how he wanted it. He kept his secret fountain to himself. Reclusive and suspicious, he was painfully sensitive to criticism and hated getting into arguments with anyone, especially those who didn't understand him. As he later put it, he didn't enjoy being "baited by little smatterers in mathematics."

He had another reason to be wary: He knew that his work could be attacked on logical grounds. He'd used algebra, not geometry, and he'd played nonchalantly with infinity, the original sin of calculus. John Wallis, whose book had so influenced Newton in his student days, had been brutally criticized for those same transgressions. Thomas Hobbes, a political philosopher and second-rate mathematician, had blasted Wallis's *Arithmetica Infinitorum* as a "scab of symbols" for its reliance on algebra and a "scurvy book" for its use of infinity. And Newton had to admit that his own work was merely analysis, not synthesis. It was good only for making discoveries, not proving them. He downplayed his infinite methods as not "worthy of public utterance" and said, many years later, "Our specious algebra is fit enough to find out, but entirely unfit to consign to writing and commit to posterity."

For these and other reasons, Newton kept his work hidden. Yet part of him wanted credit for it. He felt torn and distressed when Nicholas Mercator published a little book about logarithms in 1668 that contained the same infinite series for the natural logarithm that Newton had discovered three years earlier. The shock and disappointment of being scooped prompted Newton to write a short manuscript in 1669 about power series and circulate it privately among a few trusted acolytes. It went far beyond logarithms. Known as *De Analysi*, its full title in English is *On Analysis by Equations Unlimited in Their Number of Terms*. In 1671, he enlarged it into his main tract on calculus, *A Treatise of the Methods of Series and Fluxions*, known as *De Methodis*, but the manuscript didn't see the light of day during his lifetime; he guarded it closely and kept it for his private use. *De Analysi* was not published until 1711; *De Methodis* appeared posthumously, in 1736. Newton's estate included five thousand pages of unpublished mathematical manuscript.

So it took the world a while to discover Isaac Newton. Within the walls of Cambridge, however, he was known as a genius. In 1669, Isaac Barrow, the first Lucasian Professor and the closest thing to a mentor that Newton ever had, stepped down and recommended that Newton be appointed to the Lucasian Chair of Mathematics.

It was an ideal post for Newton. For the first time in his life, he was financially secure. The position required little teaching. He had no graduate students, and his lectures to undergraduates were poorly attended, which was just as well. The students didn't understand him anyway. They didn't know what to make of the strange, gaunt, monkish figure in his scarlet robes, with his grim face and silvery shoulder-length hair.

After Newton completed his work on *De Methodis*, his mind was as febrile as ever, but calculus was no longer his main interest. He was now deep into biblical prophecy and chronology, optics and alchemy, splitting light into colors with prisms, experimenting with mercury, sniffing his chemicals and sometimes tasting them, stoking his tin furnace day and night as he tried to turn lead into gold. Like Archimedes, he neglected his food and his sleep. He was looking for the secrets of the universe, and he had no patience for distractions.

A distraction came one day in 1676 in the form of a letter from Paris. It was from someone named Leibniz. He had a few questions about power series.

8 Fictions of the Mind

How had leibniz heard about Newton's unpublished work? It wasn't difficult. Word of Newton's discoveries had been leaking out for years. In 1669, Isaac Barrow, hoping to promote his young protégé, had sent an anonymous copy of *De Analysi* to a man named John Collins, a mathematical wannabe and impresario. Collins had put himself at the hub of a correspondence network involving British and Continental mathematicians. He was floored by the results in *De Analysi* and asked Barrow who its author was. With Newton's permission, Barrow unmasked him: "I am glad my friends paper giveth you so much satisfaction. His name is Mr. Newton; a fellow of our College, & very young . . . but of an extraordinary genius and proficiency in these things."

Collins was never someone to keep a secret in confidence. He teased his correspondents with snippets of *De Analysi* and wowed them with Newton's results without explaining where they came from. In 1675 he showed Newton's power series for the inverse sine and sine functions to a Danish mathematician named Georg Mohr, and he in turn told Leibniz about them. Leibniz sent a request to the secretary of the Royal Society of London, a German-born schmoozer and promoter of science named Henry Oldenburg: "Since I say, he [Mohr] has brought us these studies which seem to me to be very ingenious, the latter series in particular having a certain rare elegance, so I shall be grateful, Illustrious Sir, if you will send me the proof."

Oldenburg passed the request along to Newton, and Newton was not pleased. Send the proof? Ha. Instead, he replied to Leibniz, through Oldenburg, with page after page of cryptic, intimidating formulas, the full armament of *De Analysi*. Outside of Newton's inner circle, no one had ever seen math like this. And for good measure, Newton stressed that the material was old hat: "I write rather shortly because these theories long ago began to be distasteful to me, to such an extent that I have now refrained from them for nearly five years."

Undeterred, Leibniz wrote back and poked Newton, hoping to extract a bit more. He was a newcomer to all this. A diplomat, logician, linguist, and philosopher, he'd only recently become interested in advanced mathematics. He'd spent time with Christiaan Huygens, the leading mathematical mind in Europe, to get up to speed on the latest developments. After just three years of study, Leibniz had already outpaced everyone on the Continent. All he needed now was to figure out what Newton knew . . . and what he was withholding.

To pry the information out of Newton, Leibniz tried a different tack. He made the mistake of trying to impress him. He produced some of his own wares—in particular, an infinite series he was proud of—and offered it to Newton, ostensibly as a gift but actually as a signal that he was worthy to receive the secret.

Newton replied through Oldenburg two months later, on October 24, 1676. He opened with flattery, calling Leibniz "very distinguished" and praising his infinite series, saying that it "leads us also to hope for very great things from him." Were these compliments meant to be taken seriously? Apparently not, for the next line burned with acid sarcasm: "The variety of ways by which the same goal is approached has given me the greater pleasure, because three methods of arriving at series of that kind had already become known to me, so that I could scarcely expect a new one to be communicated to us." In other words, *Thanks for showing me something I already know how to do three other ways.*

In the rest of his letter, Newton toyed with Leibniz. He revealed some of his own methods for infinite series, explaining them in the pedagogical manner one would use to lecture a schoolchild. Fortunately for posterity, these parts of the letter are so clear that we can understand exactly what Newton had in mind.

But when he got to his most prized possessions (the revolutionary techniques of his second tract on calculus, *De Methodis*, including the fundamental theorem, which hadn't leaked out yet), Newton's gentle exposition came to a halt: "The foundation of these operations is evident enough, in fact; but because I cannot proceed with the explanation of it now, I have preferred to conceal it thus: 6accdae13eff7i3l9n4o4qrr4s8t12vx. On this foundation I have also tried to simplify the theories which concern the squaring of curves, and I have arrived at certain general theorems."

And with that encrypted code, Newton dangled his most cherished secret in front of Leibniz, essentially telling him, *I know something you don't, and even if you discover it later, this cryptogram will prove I knew it first.*

What Newton did not realize was that Leibniz had already discovered the secret on his own.

In the Twinkling of an Eyelid

Between 1672 and 1676, Leibniz had created his own version of calculus. Like Newton, he spotted and proved the fundamental theorem, recognized its significance, and built an algorithmic system around it. With its help, he wrote, he'd been able to derive "in the twinkling of an eyelid" nearly all the theorems about quadratures and tangents known at that time—except for the ones Newton was still hiding from the world.

When Leibniz wrote his two letters to Newton in 1676, nosing around and asking for proofs, he knew he was being pushy but he couldn't help it. As he once told a friend, "I feel myself burdened with a deficiency that counts for a great deal in this world, namely, that I lack polished manners and thereby often spoil the first impression of my person."

Skinny, stooped, and pale, Leibniz might not have been much to look at, but his mind was beautiful. He was the most versatile genius in a century of geniuses that included Descartes, Galileo, Newton, and Bach.

Although Leibniz found his calculus a decade after Newton did, he is generally considered its co-inventor for several reasons. He published it first, in a graceful and digestible form, and he couched it in a carefully designed, elegant notation that's still used today. Moreover, he attracted disciples who spread the word with evangelical zeal. They wrote influential textbooks and developed the subject in luxuriant detail. Much later, when Leibniz was accused of stealing calculus from Newton, his disciples defended him vigorously and counterattacked Newton with equal fervor.

Leibniz's approach to calculus is more elementary—and, in some ways, more intuitive—than Newton's. It also explains why the study of derivatives has long been called *differential* calculus and why the operation of taking a derivative is called *differentiation*—it's because, in Leibniz's approach, concepts called differentials are the true heart of calculus; derivatives are secondary, an afterthought, a later refinement.

Nowadays, we tend to forget how important differentials were. Modern textbooks downplay them, redefine them, or whitewash them away because they are (gasp!) infinitesimals. As such, they are seen as paradoxical, transgressive, and scary, so just to be on the safe side, many books keep infinitesimals locked in the cellar, like Norman Bates's mother in *Psycho*. But they're really nothing to be afraid of. Really.

Let's go meet Mother.

Infinitesimals

An infinitesimal is a hazy thing. It is supposed to be the tiniest number you can possibly imagine that isn't actually zero. More succinctly, an infinitesimal is smaller than everything but greater than nothing.

Even more paradoxically, infinitesimals come in different sizes. An infinitesimal part of an infinitesimal is incomparably smaller still. We could call it a second-order infinitesimal.

Just as there are infinitesimal numbers, there are infinitesimal lengths and infinitesimal times. An infinitesimal length is not a point—it's bigger than that—but it is smaller than any length you can envision. Likewise, an infinitesimal time interval is not an instant, not a single point in time, but it is shorter than any conceivable duration.

The concept of infinitesimals arose as a way of speaking about limits. Recall the example in chapter 1 where we looked at a sequence of regular polygons starting with an equilateral triangle and a square and proceeding upward through pentagons, hexagons, and other regular polygons having more and more sides. We noticed that the more sides we considered and the shorter we made them, the more the polygon began to look like a circle. We were tempted to say that a circle is an infinite polygon having infinitesimal sides but bit our tongues because the notion seemed to lead to nonsense.

We also found that if we chose any point on the circumference of the circle and looked at it under a microscope, any tiny arc containing that point looked straighter and straighter as the magnification increased. In the limit of infinite magnification, that tiny arc looks perfectly straight. In that sense, it really does seem helpful to think of the circle as an infinite collection of straight pieces and therefore as an infinite polygon with infinitesimal sides.

Both Newton and Leibniz used infinitesimals, but while Newton later disavowed them in favor of fluxions (which are ratios of first-order infinitesimals and hence finite and presentable, just like derivatives), Leibniz took a more pragmatic view. He didn't fret about whether they actually existed. He saw them as useful shorthand, an efficient way to recast arguments about limits. He also regarded them as helpful bookkeeping devices that freed the imagination for more productive work. As he explained to a colleague, "Philosophically speaking, I no more believe in infinitely small quantities than in infinitely great ones, that is, in infinitesimals rather than infinituples. I consider both as fictions of the mind for succinct ways of speaking, appropriate to the calculus."

And what do mathematicians think today? Do infinitesimals really exist? It depends on what you mean by *really*. Physicists tell us infinitesimals don't exist in the real world (but then again, neither does the rest of mathematics). Within the ideal world of mathematics, infinitesimals don't exist in the real number system, but they do exist in certain nonstandard number systems that generalize the real numbers. For Leibniz and his followers, they existed as fictions of the mind that came in handy. That's the way we will be thinking about them.

The Cube of Numbers near 2

To see how illuminating infinitesimals can be, let's start very concretely. Consider this arithmetic problem: What's 2 cubed (meaning $2 \times 2 \times 2$)? It's 8, of course. What about $2.001 \times 2.001 \times 2.001$? Slightly more than 8, sure, but how much more?

What we are after here is a way of thinking, not a numerical answer. The general question is, when we change the input to a problem (here, by changing 2 to 2.001), how much does the output change? (Here, it changes from 8 to 8 plus something whose structure we want to understand.)

Since it's hard to resist peeking, let's go ahead and see what a calculator has to say. Punching in 2.001 and hitting the x^3 button gives

$$(2.001)^3 = 8.012006001.$$

The structure to notice is that the extra bit after the decimal point is really three extra bits of very different sizes:

$$.012006001 = .012 + .000006 + .000000001.$$

Think of this as small plus super-small plus super-super-small.

We can understand the structure we're seeing by working with algebra. Suppose a quantity x (played here by the number 2) changes slightly to $x + \Delta x$ (in this case, becoming 2.001). The symbol Δx denotes the *difference* in x, meaning a tiny change in x (here, $\Delta x = 0.001$). Then, when we ask what $(2.001)^3$ is, we are really asking what $(x + \Delta x)^3$ is. Multiplying it out (or using Pascal's triangle or the binomial theorem), we find that

$$(x + \Delta x)^3 = x^3 + 3x^2\Delta x + 3x(\Delta x)^2 + (\Delta x)^3.$$

For our problem where $x = 2$, this equation becomes

$$\begin{aligned}(2 + \Delta x)^3 &= 2^3 + 3(2)^2(\Delta x) + 3(2)(\Delta x)^2 + (\Delta x)^3 \\ &= 8 + 12\Delta x + 6(\Delta x)^2 + (\Delta x)^3.\end{aligned}$$

Now we see why the extra bit beyond 8 consists of three bits of different sizes. The small but dominant bit is $12\Delta x = 12(.001) = .012$. The remaining bits $6(\Delta x)^2$ and $(\Delta x)^3$ account for the super-small .000006 and the super-super-small .000000001. The more factors of Δx there are in a bit, the smaller it is. That's why the bits are graded in size. Every additional multiplication by the tiny factor Δx makes a small bit even smaller.

The key insight behind differential calculus is displayed right here in this humble example. In many problems of cause and effect, dose and response, input and output, or any other sort of relationship between a variable x and another variable y that depends on it, a small change in the input, Δx, produces a small change in the output, Δy. That small change is typically organized in a structured way we can exploit—namely, the change in the output consists of a hierarchy of bits. They are graded in size from small to super-small to even smaller contributions. That gradation allows us to focus on the small but dominant change and neglect all the rest, the super-small and even smaller ones. That's the key insight. Although the small change is small, it is gigantic compared to the others (much like .012 was gigantic compared to .000006 and .000000001).

Differentials

This way of thinking, in which we neglect all contributions to the right answer except for the biggest one, the lion's share, might seem only approximate. And it is—if the changes in the input, like the .001 we tacked onto the 2 above, are *finite* changes. But if we consider *infinitesimal* changes in the input, then our thinking becomes exact. We make no error whatsoever. The lion's share becomes everything. And, as we've seen throughout this book, infinitesimal changes are precisely what we need to make sense of slopes, instantaneous velocities, and the areas of curved regions.

To see how this works in practice, let's go back to the example above, where we were trying to calculate the cube of a number slightly greater than 2. Except now, let's change 2 to $2 + dx$, where dx is supposed to represent an infinitesimally small difference Δx. This notion is inherently nonsensical so don't think about it too hard. The point is that learning how to work with it makes calculus a breeze.

In particular, the earlier calculation of $(2 + \Delta x)^3$ as $8 + 12\,\Delta x + 6(\Delta x)^2 + (\Delta x)^3$ now shrinks to something much simpler:

$$(2 + dx)^3 = 8 + 12\,dx.$$

What happened to the other terms like $6(dx)^2 + (dx)^3$? We discarded them. They are negligible. They are super-small and super-super-small infinitesimals and are utterly inconsequential compared to $12\,dx$. But then why do we keep $12\,dx$? Isn't it equally negligible compared to 8? It is, but if we were to discard it too, we wouldn't be considering any change at all. Our answer would be frozen at 8. So the recipe is this: to study infinitesimal change, keep terms that involve dx to the first power and ignore the rest.

This way of thinking, using infinitesimals like dx, can be rephrased in terms of limits and made perfectly kosher and rigorous. That's how modern textbooks deal with them. But it's easier and faster to use infinitesimals. The term of art for them in this context is *differentials*. Their name comes from thinking of them as being like the differences Δx and Δy, in the limit as those differences tend to zero. They are like what we saw when we looked at a parabola under a microscope and watched the curve get straighter and straighter as we zoomed in on it.

Derivatives via Differentials

Let me show you how easy certain ideas become when couched in differentials. For example, what's the slope of a curve when it's viewed as a graph in the *xy* plane? As we learned from our work with the parabola in chapter 6, the slope is the derivative of *y*, defined as the limit of $\Delta y/\Delta x$ as Δx approaches zero. But what is it in terms of differentials? It's simply dy/dx. It's as if the curve is made up of little straight pieces:

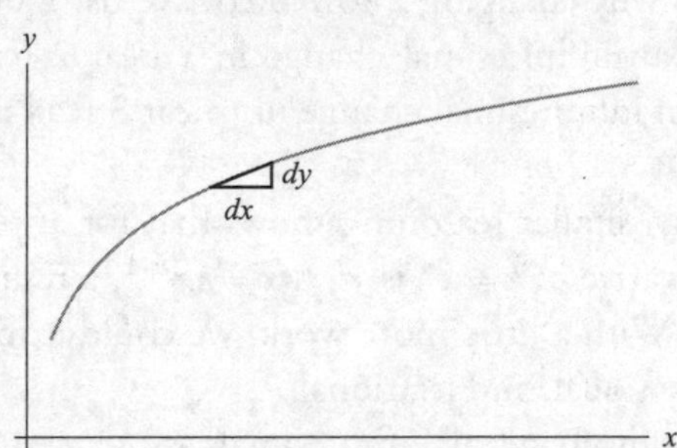

If we think of *dy* as an infinitesimal rise and *dx* as an infinitesimal run, the slope is simply the rise over the run, just as it always is, and hence is dy/dx.

To apply this approach to a specific curve (say $y = x^3$, the case we considered while cubing numbers slightly greater than 2), we calculate *dy* as follows. Write

$$y + dy = (x + dx)^3.$$

As before, the right-hand side expands to

$$(x + dx)^3 = x^3 + 3x^2dx + 3x(dx)^2 + (dx)^3.$$

But now, following the recipe, we discard the terms $(dx)^2$ and $(dx)^3$, since they're not part of the lion's share. Thus

$$y + dy = (x + dx)^3 = x^3 + 3x^2dx.$$

And since $y = x^3$, we can simplify the equation above to obtain

$$dy = 3x^2dx.$$

Dividing both sides by *dx* yields the corresponding slope,

$$\frac{dy}{dx} = 3x^2.$$

At $x = 2$, this gives a slope of $3(2)^2 = 12$. That's the same 12 we saw earlier. It's why changing 2 to 2.001 gave us $(2.001)^3 \approx 8.012$. It means that an infinitesimal change in x near 2 (call it dx) gets converted to an infinitesimal change in y near 8 (call it dy) that's 12 times bigger ($dy = 12dx$).

Incidentally, similar reasoning shows that for any positive integer n, the derivative of $y = x^n$ is $dy/dx = nx^{n-1}$, a result we've mentioned earlier. With a little more work, we could extend this result to negative, fractional, and irrational n.

The great advantage of infinitesimals in general and differentials in particular is that they make calculations easier. They provide shortcuts. They free the mind for more imaginative thought, just as algebra did for geometry in an earlier era. This is what Leibniz adored about his differentials. As he wrote to his mentor Huygens, "My calculus gave me, almost without meditation, the great part of the discoveries which have been made concerning this subject. For what I love most about my calculus is that it gives us the same advantages over the Ancients in the geometry of Archimedes, that Viète and Descartes have given us in the geometry of Euclid or Apollonius, in freeing us from having to work with the imagination."

The only thing wrong with infinitesimals is that they don't exist, at least not within the system of real numbers. Oh, and one other thing—they are paradoxical. They wouldn't make sense even if they did exist. One of Leibniz's disciples, Johann Bernoulli, realized they'd have to satisfy nonsensical equations like $x + dx = x$, even though dx isn't zero. Hmmm. Well, you can't have everything. Infinitesimals do give the right answers once we learn how to work with them, and the benefits they provide more than make up for any psychic distress they may cause. They are like Picasso's lie that helps us realize the truth.

As a further demonstration of infinitesimals' power, Leibniz used them to derive Snell's sine law for the refraction of light. Recall from chapter 4 that when light passes from one medium into another—let's say from air into water—it bends in accordance with a mathematical law that was discovered and rediscovered several times over the centuries. Fermat had explained it with his principle of least time, but he struggled mightily to solve the optimization problem that his principle implied. With his new calculus of differentials, Leibniz deduced the sine law with ease and noted with evident pride that "other very learned men have sought in many devious ways what someone versed in this calculus can accomplish in these lines as by magic."

The Fundamental Theorem via Differentials

Another triumph of Leibniz's differentials is that they made the fundamental theorem transparent. Recall that the fundamental theorem concerns the area accumulation function $A(x)$, which gives the area under the curve $y = f(x)$ over the interval from 0 to x. The theorem says that as we slide x to the right, the area under the curve accumulates at a rate given by $f(x)$ itself. Thus $f(x)$ is the derivative of $A(x)$.

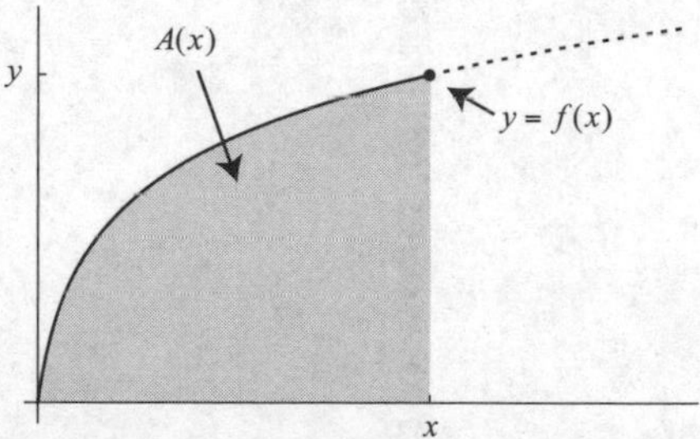

To see where this result comes from, suppose we change x by an infinitesimal amount to $x + dx$. How much does the area $A(x)$ change? By definition, it changes by an amount dA. Hence the new area equals the old area plus the change in area and is therefore $A + dA$.

The fundamental theorem drops out immediately once we visualize what dA must be. As suggested by the picture below, the area changes by the infinitesimal amount dA given by the area of the infinitesimally thin vertical strip between x and $x + dx$:

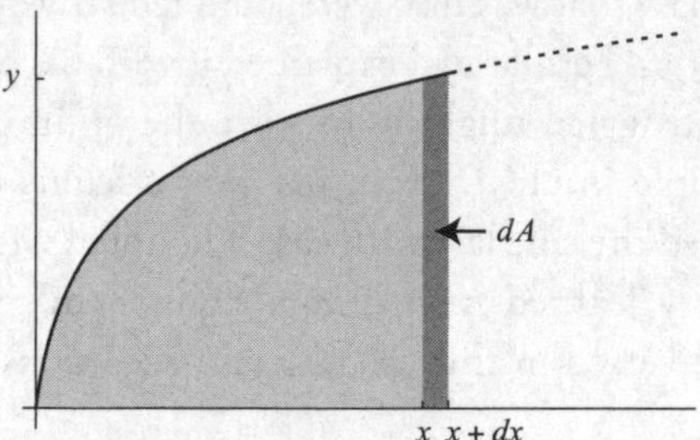

That strip is a rectangle of height y and base dx. So its area is its height times its base, which is $y\ dx$ or, if you prefer, $f(x)\ dx$.

Actually, the strip is a rectangle only when viewed infinitesimally. In reality, for a strip of any finite width Δx, the change in area ΔA has two contributions. The dominant one is a rectangle of area $y\ \Delta x$. A much smaller one is the area of the tiny, curved, triangular-looking cap on top of the rectangle.

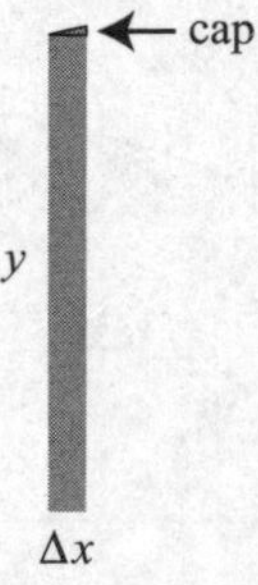

Here's another case where the infinitesimal world is nicer than the real world. In the real world, we would have to account for the area of the cap, which wouldn't be easy to estimate because it would depend on the details of the curve on top. But as the width of the rectangle approaches zero and "becomes" dx, the area of the cap becomes negligible compared to the area of the rectangle. It's super-small compared to small.

The upshot is that $dA = y\,dx = f(x)\,dx$. Boom—that's the fundamental theorem of calculus. Or, as it is more politely phrased nowadays (in our misguided era when differentials have been forsaken for derivatives),

$$\frac{dA}{dx} = y = f(x).$$

This is exactly what we found in chapter 7 with the paint-roller argument.

One last thing: When we regard the area under a curve as a sum of infinitely many infinitesimal rectangular strips, we write it as

$$A(x) = \int_0^x f(x)\,dx.$$

That long-necked, swan-like symbol is actually a stretched-out *S*. The *S* reminds us that a summation is taking place. It's a summation of a peculiar kind, distinctive to integral calculus, involving a sum of infinitely many infinitesimal strips, all being integrated into a single, coherent area. As a symbol of integration, it's called an integral sign. Leibniz introduced it in a 1677 manuscript and published it in 1686. It's calculus's most recognizable icon. The zero at the bottom of it and the *x* at the top of it indicate the endpoints of the interval of the *x*-axis over which the rectangles stand. Those endpoints are called the limits of integration.

What Led Leibniz to Differentials and the Fundamental Theorem?

Newton and Leibniz arrived at the fundamental theorem of calculus by two separate routes. Newton came at it by thinking about motion and flow, the continuous side of math. Leibniz came at it from the other side. Although he was not a mathematician by training, earlier in his life he'd spent some time thinking about discrete math

—whole numbers and counting, combinations and permutations, and fractions and sums of a particular sort.

He began wading into deeper waters after he met Christiaan Huygens. At the time, Leibniz was serving on a diplomatic mission in Paris, but he found himself entranced by what Huygens was telling him about the latest developments in mathematics and he wanted to learn more. With amazing pedagogical prescience (or was it luck?), Huygens challenged his student with a problem that led him to the fundamental theorem.

The problem he gave him was to calculate this infinite sum:

$$\frac{1}{1\cdot 2}+\frac{1}{2\cdot 3}+\frac{1}{3\cdot 4}+\cdots+\frac{1}{n\cdot(n+1)}+\cdots=?$$

(The dots in the denominators mean multiplication.) To bring the problem down to earth, let's begin with a warm-up version. Suppose the sum has, say, 99 terms in it instead of infinitely many. Then we would have to calculate

$$S=\frac{1}{1\cdot 2}+\frac{1}{2\cdot 3}+\frac{1}{3\cdot 4}+\cdots+\frac{1}{n\cdot(n+1)}+\cdots+\frac{1}{99\cdot 100}.$$

If you don't see the trick, this is a tedious but straightforward calculation. With sufficient patience (or a computer), we could ploddingly add up the 99 fractions. But that would be missing the point. The point is to find an *elegant* solution. Elegant solutions are valued in math in part because they're pretty but also because they're powerful. The light they shed can often be used to illuminate *other* problems. In this case, the elegant light that Leibniz discovered quickly pointed him to the fundamental theorem.

He solved Huygens's problem with a brilliant trick. The first time I saw it, I felt like I was watching a magician pull a rabbit out of a hat. If you want to experience that same feeling, skip the analogy I'm about to present. But if you prefer to understand what's behind the magic, here's what makes it work.

Imagine someone climbing a very long and irregular staircase.

Suppose our climber wanted to measure the total vertical rise from the bottom of the staircase to the top. How could he do that? Well, he could always add up all the rises of the individual steps in between. That uninspired strategy would be like adding up the 99 terms in the sum S above. It could be done, but it would be unpleasant because the staircase is so irregular. And if the staircase has millions of steps, adding up all their rises would be a hopeless task. There has to be a better way.

The better way is to use an *altimeter*. An altimeter is a device that measures altitude. If Zeno in the picture had an altimeter, he could solve his problem by subtracting the altitude at the bottom of the staircase from the altitude at the top. That's all there is to it: the total vertical rise equals the difference in those two altitudes. The difference between them has to equal the sum of all the rises in between. No matter how irregular the staircase is, this trick will always work.

The success of the trick hinges on the fact that the altimeter readings are intimately related to the rises of the steps—the rise of any given step is the difference of consecutive altimeter readings. In other words, the height of a step equals the altitude at its top minus the altitude at its bottom.

By now you're probably thinking, *What does an altimeter have to do with the original math problem of adding up a long list of complicated, irregular numbers?* Well, if we could somehow find the analog of an altimeter for a complicated, irregular sum, that sum would become easy. It would just add up to the difference between the highest and lowest altimeter readings. This is essentially what Leibniz did. He found an altimeter for the sum S. It enabled him to write each term in the sum as a difference of consecutive altimeter readings, which in turn allowed him to compute the desired sum using the idea mentioned above. Then he generalized his altimeter to other problems. Ultimately it led him to the fundamental theorem of calculus.

With this analogy in mind, let's examine S again:

$$S=\frac{1}{1\cdot 2}+\frac{1}{2\cdot 3}+\frac{1}{3\cdot 4}+\cdots+\frac{1}{n\cdot(n+1)}+\cdots+\frac{1}{99\cdot 100}.$$

We're going to rewrite each term as a difference of two other numbers. This is like saying that the rise of each step is the difference of the altimeter readings at its top and bottom. For the first step, the rewriting goes like this:

$$\frac{1}{1\cdot 2}=\frac{2-1}{1\cdot 2}=\frac{1}{1}-\frac{1}{2}.$$

Admittedly, it's not obvious yet where this is going, but stay tuned. In a moment we'll see how helpful it is to rewrite the fraction $1/(1\cdot 2)$ as a difference of two consecutive unit fractions, $1/1$ and $1/2$. (A *unit fraction* means a fraction with a 1 in the numerator. These consecutive unit fractions are going to play the role of consecutive altimeter readings.) Also, if the arithmetic above seems unclear, try simplifying the equations by working them from right to left. On the far right we are subtracting a unit fraction ($1/2$) from another unit fraction ($1/1$); in the middle we are putting them over a common denominator; and on the far left we are simplifying the numerator.

Similarly, we can write every other term in S as a difference of consecutive unit fractions:

$$\frac{1}{2\cdot 3}=\frac{3-2}{2\cdot 3}=\frac{1}{2}-\frac{1}{3}$$

$$\frac{1}{3\cdot 4}=\frac{4-3}{3\cdot 4}=\frac{1}{3}-\frac{1}{4}$$

and so on. When we add up all these differences of unit fractions, S becomes

$$S=\left(\tfrac{1}{1}-\tfrac{1}{2}\right)+\left(\tfrac{1}{2}-\tfrac{1}{3}\right)+\left(\tfrac{1}{3}-\tfrac{1}{4}\right)+\cdots+\left(\tfrac{1}{98}-\tfrac{1}{99}\right)+\left(\tfrac{1}{99}-\tfrac{1}{100}\right).$$

Now we see the method in the madness. Look carefully at the structure of this sum. Nearly all the unit fractions appear twice, once with a negative sign and once with a positive sign. For example, ½ is subtracted and then added back in; the net effect is that the ½ terms cancel each other out. The same is true for ⅓. It occurs twice and cancels itself. Nearly all the other unit fractions, up to and including 1⁄99, do the same. The only exceptions are the first and last unit fractions, 1⁄1 and 1⁄100. Being at the ends of the line in S, they have no partners to cancel with. After the smoke clears, they are the only unit fractions left standing. So the result is

$$S=\tfrac{1}{1}-\tfrac{1}{100}.$$

This makes perfect sense in terms of the staircase analogy. It says the total rise of all the steps is the altitude at the top of the staircase minus the altitude at the bottom.

Incidentally, S simplifies to 99⁄100. That's the answer to the puzzle with 99 terms. Leibniz realized that he could add *any* number of terms using the same trick. If the sum had N terms instead of 99, the result would be

$$S=\tfrac{1}{1}-\tfrac{1}{N+1}.$$

Thus the answer to Huygens's original question about the infinite sum becomes clear: As N approaches infinity, the term $1/(N+1)$ approaches zero, and so S approaches 1. That limiting value of 1 is the answer to Huygens's puzzle.

The key that allowed Leibniz to find the sum was that it had a very particular structure: it could be rewritten as a sum of consecutive differences (in this case, differences of consecutive unit fractions). That difference structure caused the massive cancellations we saw above. Sums with this property are now termed *telescoping sums* because they call to mind one of those old collapsible telescopes you see in pirate movies, the kind of spyglass that can be stretched out or contracted at will. The analogy is that the original sum appears in its stretched-out form, but, because of its difference structure, it can be telescoped down to a much more compact result. The only terms that survive the collapse are the terms without partners to cancel them, the ones at the very ends of the telescope.

Naturally Leibniz wondered if he could use the telescoping trick on other problems. It was an idea worth pursuing, given how powerful it could be. Confronted with a long list of numbers to sum, if he could write each number as a difference of consecutive numbers (to be determined), the telescoping trick would work again.

And that got Leibniz thinking about areas. Approximating the area under a curve in the *xy* plane, after all, amounted to summing a long list of numbers, the areas of lots of thin, vertical rectangular strips.

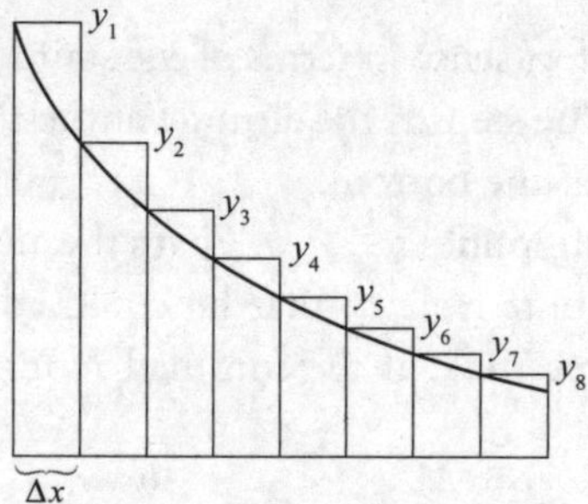

The idea behind what he had in mind is demonstrated in the figure above. It shows only eight rectangular areas, but you should try to imagine a similar image with millions or billions of much thinner rectangles or, better yet, infinitely many infinitesimally thin rectangles. That, unfortunately, is hard to draw or visualize. That's why I'm using eight blocky rectangles for now.

For simplicity, suppose all the rectangles have the same width. Call it Δx. The heights of the rectangles are $y_1, y_2, \ldots y_8$. Then the total area of the approximating rectangles is

$$y_1\Delta x + y_2\Delta x + \cdots + y_8\Delta x.$$

This sum of eight numbers would conveniently telescope if we could somehow find magic numbers $A_0, A_1, A_2, \ldots, A_8$ whose differences give the rectangular areas

$$y_1\Delta x = A_1 - A_0$$
$$y_2\Delta x = A_2 - A_1$$
$$y_3\Delta x = A_3 - A_2$$

and so on, down to $y_8\Delta x = A_8 - A_7$. Then the total area of the rectangles would telescope to this:

$$\begin{aligned} y_1\Delta x + y_2\Delta x + \cdots + y_8\Delta x &= (A_1 - A_0) + (A_2 - A_1) + \cdots + (A_8 - A_7) \\ &= A_8 - A_0. \end{aligned}$$

Now think about the limit of infinitesimally thin strips. Their width Δx turns into the differential dx. Their varying heights $y_1, y_2, \ldots, y_8$ become $y(x)$, a function that gives the height of the rectangle standing over the point labeled by the variable x. The sum of the infinitely many rectangular areas becomes the integral $\int y(x)\,dx$. And as for the earlier telescoping, the sum that was previously $A_8 - A_0$ now becomes $A(b) - A(a)$, where a and b are the values of x on the left and right ends of the area being calculated. The infinitesimal version of telescoping then yields the *exact* area under the curve:

$$\int_a^b y(x)\,dx = A(b) - A(a).$$

And how do we find the magic function $A(x)$ that makes all this possible? Well, look at the earlier equations like $y_1\Delta x = A_1 - A_0$. They morph into

$$y(x)\,dx = dA$$

as the rectangles become infinitesimally thin. To put the same result in terms of derivatives instead of differentials, divide both sides of the equation above by dx to get

$$\frac{dA}{dx} = y(x).$$

This is how we find the analogs of the magic numbers $A_0, A_1, A_2, \ldots, A_8$ that cause telescoping to occur. In the limit of infinitesimally thin strips, they are given by the unknown function $A(x)$ whose derivative is the given curve $y(x)$.

All of this is Leibniz's version of the backward problem and the fundamental theorem of calculus. As he put it, "Finding the areas of figures is reduced to this: given a series, to find the sums, or (to explain this better) given a series, to find another one whose differences coincide with the terms of the given series." In this way, differences and telescoping sums guided Leibniz to differentials and integrals and from there to the fundamental theorem, just as fluxions and expanding areas had led Newton to that same secret fountain.

Fighting HIV with an Assist from Calculus

Although differentials are fictions of the mind, they have affected our world, our societies, and our lives in profoundly nonfictional ways ever since Leibniz invented them. For an example in our own time, consider the supporting role that differentials played in the understanding and treatment of HIV, the human immunodeficiency virus.

In the 1980s, a mysterious disease began killing tens of thousands of people a year in the United States and hundreds of thousands worldwide. No one knew what it was, where it came from, or what was causing it, but its effects were clear—it weakened patients' immune systems so severely that they became vulnerable to rare kinds of cancer, pneumonia, and opportunistic infections. Death from the disease was slow, painful, and disfiguring. Doctors named it acquired immune deficiency syndrome, or AIDS. Patients and doctors were desperate. No cure was in sight.

Basic research demonstrated that a retrovirus was the culprit.

Its mechanism was insidious: The virus attacked and infected white blood cells called helper T cells, a key component of the immune system. Once inside, the virus hijacked the cell's genetic machinery and co-opted it into making more viruses. Those new virus particles then escaped from the cell, hitched a ride in the bloodstream and other bodily fluids, and looked for more T cells to infect. The body's immune system responded to this invasion by trying to flush out the virus particles from the blood and kill as many infected T cells as it could find. In so doing, the immune system was killing an important part of itself.

The first antiretroviral drug approved to treat HIV appeared in 1987. Although it slowed HIV down by interfering with the hijacking process, it wasn't as effective as hoped, and the virus often became resistant to it. A different class of drugs called protease inhibitors appeared in 1994. They thwarted HIV by interfering with the newly produced virus particles, keeping them from maturing and rendering them noninfectious. Although also not a cure, protease inhibitors were a godsend.

Soon after protease inhibitors became available, a team of researchers led by Dr. David Ho (a former physics major at Caltech and so, presumably, someone comfortable with calculus) and a mathematical immunologist named Alan Perelson collaborated on a study that changed how doctors thought about HIV and revolutionized how they treated it. Before the work of Ho and Perelson, it was known that untreated HIV infection typically progressed through three stages: an acute primary stage of a few weeks, a chronic and paradoxically asymptomatic stage of up to ten years, and a terminal stage of AIDS.

In the first stage, soon after a person becomes infected with HIV, he or she displays flu-like symptoms of fever, rash, and headaches, and the number of helper T cells (also known as CD4 cells) in the bloodstream plummets. A normal T-cell count is about 1000 cells per cubic millimeter of blood; after primary HIV infection, T-cell count drops to the low hundreds. Since T cells help the body fight infections, their depletion severely weakens the immune system. Meanwhile, the number of virus particles in the blood, known as the viral load, spikes and then drops as the immune system begins to combat the HIV infection. The flu-like symptoms disappear and the patient feels better.

At the end of this first stage, the viral load stabilizes at a level that can, puzzlingly, last for many years. Doctors refer to this level as the set point. A patient who is untreated may survive for a decade with no HIV-related symptoms and no lab findings other than a persistent viral load and a low and slowly declining T-cell count. Eventually, however, the asymptomatic stage ends and AIDS sets in, marked by a further decrease in the T-cell count and a sharp rise in the viral load. Once an untreated patient has full-blown AIDS, opportunistic infections, cancers, and other complications usually cause the patient's death within two to three years.

The key to the mystery was in the decade-long asymptomatic stage. What was going on then? Was HIV lying dormant in the body? Other viruses were known to hibernate like that. The genital-herpes virus, for example, hunkers down in nerve ganglia to evade the immune system. The chickenpox virus also does this, hiding out in nerve cells for years and sometimes awakening to cause shingles. For HIV, the reason for the latency was unknown, but it became clear after Ho and Perelson's work.

In a 1995 study, they gave patients a protease inhibitor, not as a treatment but as a probe. This nudged a patient's body off its set point and allowed Ho and Perelson—for the first time ever—to track the dynamics of the immune system as it battled HIV. They found that after each patient took the protease inhibitor, the number of virus particles in his bloodstream dropped exponentially fast. The rate of decay was incredible; half of all the virus particles in the bloodstream were cleared by the immune system every *two days*.

Differential calculus enabled Perelson and Ho to model this exponential decay and extract its surprising implications. First they represented the changing concentration of virus in the blood as an unknown function, $V(t)$, where t denotes the elapsed time since the protease inhibitor was administered. Then they hypothesized how much the concentration of virus would change, dV, in an infinitesimally short time interval, dt. Their data indicated that a constant fraction of the virus in the blood was cleared each day, so perhaps the same constancy would hold when extrapolated down to an infinitesimal time interval dt. Since dV/V is the fractional change in the virus concentration, their model could be translated into symbols as the following equation:

$$\frac{dV}{V} = -c\,dt.$$

Here the constant of proportionality, c, is the clearance rate, a measure of how fast the body flushed out the virus.

The equation above is an example of a *differential equation*. It relates the differential dV to V itself and to the differential dt of the elapsed time. By using the fundamental theorem to integrate both sides of the equation, Perelson and Ho solved for $V(t)$ and found it satisfied

$$\ln[V(t)/V_0] = -ct$$

where V_0 is the initial viral load and ln denotes the natural logarithm (the same logarithmic function that Newton and Mercator studied in the 1660s). Inverting this function then implied

$$V(t) = V_0 e^{-ct},$$

where e is the base of the natural logarithm, thus confirming that the viral load did indeed decay exponentially fast in the model. Finally, by fitting an exponential-decay curve to their experimental data, Ho and Perelson estimated the previously unknown value of the clearance rate c.

For those who prefer derivatives to differentials, the model equation can be rewritten as

$$\frac{dV}{dt} = -cV.$$

Here, dV/dt is the derivative of V. It measures how fast the virus concentration grows or declines. Positive values of the derivative signify growth; negative values indicate decline. Since the concentration V is positive, then $-cV$ must be negative, so the derivative must also be negative, which means the virus concentration must decline, as we know it does in the experiment. Furthermore, the proportionality between dV/dt and V means that the closer V gets to zero, the more slowly it declines. Intuitively, this slowing decline of V is like what happens if you fill a sink with water and then allow it to drain. The less water there is in the sink, the more slowly it flows out because there's less water pressure pushing it down. In this analogy, the amount of virus is like the water, and the draining is like the outflow of the virus due to its clearance by the immune system.

Having modeled the effect of the protease inhibitor, Perelson and Ho modified their equation to describe the conditions *before* the drug was administered. They assumed the equation would become

$$\frac{dV}{dt} = P - cV.$$

In this equation, P refers to the uninhibited rate of production of new virus particles, another crucial unknown at that time. Perelson and Ho imagined that before administration of the protease inhibitor, at every moment infected cells were releasing new infectious virus particles, which then infected other cells, and so on. This potential for a raging fire is what makes HIV so devastating.

In the asymptomatic phase, however, there is evidently a balance between the production of the virus and its clearance by the immune system. At this set point, the virus is produced as fast as it's cleared. That gave new insight into why the viral load could stay the same for years. In the water-in-the-sink analogy, it's like what happens if you turn on the faucet and open the drain at the same time. The water will reach a steady-state level at which outflow equals inflow.

At the set point, the concentration of virus doesn't change, so its derivative must be zero: $dV/dt = 0$. Hence, the steady-state viral load, V_0, satisfies

$$P = cV_0.$$

Perelson and Ho used this simple equation, $P = cV_0$, to estimate a vitally important number that no one had found a way to measure before: the number of virus particles being cleared each day by the immune system. It turned out to be a *billion* virus particles a day.

That number was unexpected and truly stunning. It indicated that a titanic struggle was taking place during the seemingly calm ten years of the asymptomatic phase in a patient's body. Every day, the immune system cleared a billion virus particles and the infected cells released a billion new ones. The immune system was in a furious, all-out war with the virus and fighting it to a near standstill.

Ho, Perelson, and their colleagues conducted a follow-up study in 1996 to get a better handle on something they'd seen in 1995 but couldn't resolve back then. This time they collected viral-load data at shorter time intervals after the protease inhibitor was administered because they wanted to obtain more information about an initial lag they'd observed in the medicine's absorption, distribution, and penetration into the target cells. After the drug was given, the team measured the patients' viral load every two hours until the sixth hour, then every six hours until day two, and then once a day thereafter until day seven. On the mathematical side, Perelson refined the differential-equation model to account for the lag and to track the dynamics of another important variable, the changing number of infected T cells.

When the researchers redid the experiment, fit the data to the model's predictions, and estimated its parameters again, they obtained results even more staggering than before: *ten billion* virus particles were being produced and then cleared from the bloodstream each day. Moreover, they found that infected T cells had a lifespan of only about two days. The surprisingly short lifespan added another piece to the puzzle, given that T-cell depletion is the hallmark of HIV infection and AIDS.

The discovery that HIV replication was so astonishingly rapid changed the way that doctors treated their HIV-positive patients. Until the work of Ho and Perelson, physicians waited until HIV emerged from its supposed hibernation before they prescribed antiviral drugs. The idea was to conserve forces until the patient's immune system really needed help, because the virus often became resistant to the drugs, and then there'd be nothing else to try. So it was generally thought wiser to wait until patients were far along in their illness.

Ho and Perelson's work turned this picture upside down. There was no hibernation. HIV and the body were locked in a pitched struggle every second of every day, and the immune system needed all the help it could get and as soon as possible after the critical early days of infection. And now it was obvious why no single medication worked for very long. The virus replicated so rapidly and mutated so quickly, it could find a way to escape almost any therapeutic drug.

Perelson's mathematics gave a quantitative estimate of how many drugs had to be used in combination to beat HIV down and keep it down. By taking into account the measured mutation rate of HIV, the size of its genome, and the newly estimated number of virus particles that were produced daily, he demonstrated mathematically that HIV was generating every possible mutation at every base in its genome many times a day. Since even a single mutation could confer drug resistance, there was little hope of success with single-drug therapy. Two drugs given at the same time would stand a better chance of working, but Perelson's calculations showed that a sizable fraction of all possible double mutations also occurred each day. Three drugs in combination, however, would be hard for the HIV virus to overcome. The math suggested that the odds were something like ten million to one against HIV being able to undergo the necessary three simultaneous mutations to escape triple-combination therapy.

When Ho and his colleagues tested a three-drug cocktail on HIV-infected patients in clinical studies, the results were remarkable. The level of virus in the blood dropped about a hundredfold in two weeks. Over the next month, it became undetectable.

This is not to say that HIV was eradicated. Studies soon afterward showed the virus could rebound aggressively if patients took a break from therapy. The problem is that HIV can hide out in various places in the body. It can lie low in sanctuary sites that the drugs cannot readily penetrate or lurk in latently infected cells and rest without replicating, a sneaky way of evading treatment. At any time, these dormant cells can wake up and start making new viruses. That's why it's so important for HIV-positive people to keep taking their meds, even when their viral loads are low or undetectable.

Still, even though it did not cure HIV, triple-combination therapy changed it to a chronic condition that could be managed, at least for those who had access to treatment. It gave hope where almost none had existed before.

In 1996, Dr. David Ho was named *Time* magazine's Man of the Year. In 2017, Alan Perelson received a major prize for his "profound contributions to theoretical immunology, which bring insight and save lives." He is still using calculus and differential equations to analyze viral dynamics. His latest work concerns hepatitis C, a virus that affects about 170 million people worldwide and kills about 350,000 people each year. It is the leading cause of cirrhosis and liver cancer. In 2014, with the help of Perelson's math, new treatments for hepatitis C were developed that are safe and easy to take as a once-a-day pill. Incredibly, the treatment cures the infection in nearly every patient.

9

The Logical Universe

Calculus underwent a metamorphosis in the second half of the seventeenth century. It became so systematic, so penetrating, and so powerful that many historians say calculus was "invented" then. According to this view, before Newton and Leibniz, there was proto-calculus; afterward, calculus. I wouldn't put it that way myself. To me, it's been calculus all along, ever since Archimedes harnessed infinity.

Whatever it's called, calculus transformed dramatically between 1664 and 1676, and it changed the world along with it. In science, it allowed humanity to start reading the book of nature that Galileo had dreamed of. In technology, it launched the industrial revolution and the information age. In philosophy and politics, it left its mark on modern conceptions of human rights, society, and laws.

I wouldn't say calculus was invented in the late seventeenth century; rather, I would describe what happened as an evolutionary breakthrough, analogous to a pivotal event in biological evolution. In the early days of life, organisms were relatively simple. They were single-celled creatures, something like the bacteria of today. That era of unicellular life continued for about three and a half billion years, dominating most of the Earth's history. But around half a billion years ago, an astonishing diversity of multicellular life burst forth in what biologists call the Cambrian explosion. In just a few tens of millions of years—an evolutionary split second—many of the major animal phyla suddenly emerged. Similarly, calculus was the Cambrian explosion for mathematics. Once it arrived, an amazing diversity of mathematical fields began to evolve. Their lineage is visible in their calculus-based names, in adjectives like *differential* and *integral* and *analytic*, as in differential geometry, integral equations, and analytic number theory. These advanced branches of mathematics are like the many branches and species of multicellular life. In this analogy, the microbes of mathematics are the earliest topics: numbers, shapes, and word problems. Like unicellular organisms, they dominated the mathematical scene for most of its history. But after the Cambrian explosion of calculus three hundred and fifty years ago, new mathematical life forms began to proliferate and flourish, and they altered the landscape around them.

Much of the story of life is a tale of progress toward greater sophistication and complexity building on earlier precursors. That's true of calculus as well. But what is the story building toward? Is there any direction to the evolution of calculus? Or is it, as some would say of biological evolution, undirected and random?

Within pure mathematics, the evolution of calculus has been a story of crossbreeding and its benefits. Older parts of math were invigorated after they were crossed with calculus. For example, the ancient study of numbers and their patterns was revitalized by an infusion of calculus-based tools like integrals, infinite sums, and power series. The resulting hybrid field is called analytic number theory. Likewise, differential geometry used calculus to shed light on the structure of smooth surfaces and revealed cousins they never knew they had, unimaginable curved shapes in four dimensions and beyond. In this way, the Cambrian explosion of calculus made mathematics more abstract and more powerful. It also made it more like a family. Calculus exposed a web of hidden relationships tying all parts of mathematics together.

In applied mathematics, the evolution of calculus has been a story of our expanding understanding of change. As we've seen, calculus began with the study of curves, where the changes were changes in direction, and it continued with the study of motion, where the changes became changes in position. In the aftermath of its Cambrian explosion, and especially with the rise of differential equations, calculus moved on to the study of change much more generally. Today, differential equations help us predict how epidemics will spread, where a hurricane will hit land, and how much to pay for an option to buy a stock in the future. In every field of human endeavor, differential equations have emerged as a common framework for describing how things change around us and inside us, from the subatomic domain to the farthest reaches of the cosmos.

The Logic of Nature

The earliest triumph of differential equations altered the course of Western culture. In 1687, Isaac Newton proposed a system of the world that demonstrated the power of reason and ushered in the Enlightenment. He discovered a small set of equations—his laws of motion and gravity—that could explain the mysterious patterns Galileo and Kepler had found in falling bodies on Earth and planetary orbits in the solar system. In so doing, he erased the distinction between the earthly and celestial realms. After Newton, there was just one universe, with the same laws applying everywhere and always.

In his magisterial three-volume masterpiece, *Mathematical Principles of Natural Philosophy* (often known as the *Principia*), Newton applied his theories to much more: the shape of the Earth, with its slightly bulging waistline caused by the centrifugal force of its spin; the rhythm of the tides; the eccentric orbits of comets; and the motion of the moon, a problem so difficult that Newton complained to his friend Edmond Halley that it had "made his head ache, and kept him awake so often, that he would think of it no more."

Today, when college students study physics, they are taught classical mechanics first—the mechanics of Newton and his successors—after which they are told that it has been superseded by Einstein's relativity theory and the quantum theory of Planck, Einstein, Bohr, Schrödinger, Heisenberg, and Dirac. There's certainly a lot of truth to that. The new theories overturned Newtonian conceptions of space and time, mass and energy, and determinism itself, replacing it in the case of quantum theory with a more probabilistic, statistical description of nature.

But what has not changed is the role of calculus. In relativity, as in quantum mechanics, the laws of nature are still written in the language of calculus, with sentences in the form of differential equations. That, to me, is Newton's greatest legacy. He showed that nature is logical. Cause and effect in the natural world behave much like a proof in geometry, with one truth following from another by logic, except that what is following is one *event* from another in the world, not one *idea* from another in our minds.

This uncanny connection between nature and mathematics harks back to the Pythagorean dream. The link between musical harmony and numbers discovered by the Pythagoreans led them to proclaim that *all* is number. They were onto something. Numbers are important to the workings of the universe. Shapes are important too; in the book of nature that Galileo dreamed of, the words were geometrical figures. But as important as numbers and shapes might be, they're not the true drivers of the play. In the drama of the universe, shapes and numbers are like actors; they are quietly directed by an unseen presence, the logic of differential equations.

Newton was the first to tap into this logic of the universe and build a system around it. It wasn't possible before him, because the necessary concepts hadn't been born yet. Archimedes didn't know about differential equations. Neither did Galileo, Kepler, Descartes, or Fermat. Leibniz did, but he wasn't as inclined toward science as Newton or nearly as virtuosic mathematically. The secret logic of the universe was vouchsafed to Newton alone.

The centerpiece of his theory is his differential equation of motion:

$$F = ma.$$

It ranks as one of the most consequential equations in history. It says that the force, F, on a moving body is equal to the body's mass, m, times its acceleration, a. It's a differential equation because acceleration is a derivative (the rate of change of the body's velocity) or, in Leibnizian terms, the ratio of two differentials:

$$a = \frac{dv}{dt}.$$

Here dv is the infinitesimal change in the body's velocity v during an infinitesimal time interval dt. So if we know the force F on the body, and if we know its mass m, we can use $F = ma$ to find its acceleration via $a = F/m$. That acceleration in turn determines how the body will move. It tells us how the body's velocity will change in that next instant, and its velocity tells us how its position will change. In this way, $F = ma$ is an oracle. It predicts the body's future behavior, one tiny step at a time.

Consider the simplest, bleakest situation imaginable: an isolated body alone in an empty universe. How would it move? Well, since there's nothing around to push it or pull it, the force on the body is zero: $F = 0$. Then, since m is not zero (assuming the body has some mass), Newton's law yields $F/m = a = 0$, which implies that $dv/dt = 0$ as well. But $dv/dt = 0$ means the lonesome body's velocity doesn't change during the infinitesimal time interval dt. Nor does it change during the next interval, or the one after that. The upshot is that when $F = 0$, a body maintains its velocity forever. This is Galileo's principle of inertia: In the absence of an outside force, a body at rest stays at rest, and a body in motion stays in motion and moves at a constant velocity. Its speed and direction never change. We have just deduced the law of inertia as a logical consequence of Newton's deeper law of motion, $F = ma$.

Newton seemed to have understood early on, back in his college days, that acceleration was proportional to force. He knew from studying Galileo that if there was no force on a body, it would either stay at rest or continue moving in a straight line at a constant speed. Force, he realized, was not needed to produce motion; it was needed to produce *changes* in motion. It was force that was responsible for making bodies speed up, slow down, or depart from a straight path.

This insight was a big advance over earlier Aristotelian thinking. Aristotle didn't appreciate inertia. He imagined that a force was needed just to keep a body moving. And to be fair, that's true in situations dominated by friction. If you're trying to slide a desk across the floor, you have to keep pushing it; once you stop pushing, the desk stops moving. But friction is much less relevant for planets gliding through space or apples dropping to the ground. In those cases, the force of friction is negligible. It can be ignored without losing the essence of the phenomenon.

In Newton's picture of the universe, the dominant force is gravity, not friction. Which is as it should be, given that Newton and gravity are so closely associated in the popular mind. When most people think of Newton, they immediately recall what they learned as children, that Newton discovered gravity when an apple fell on his head. Spoiler alert: That's not what happened. Newton didn't discover gravity; people already knew that heavy things fell. But nobody knew how far gravity went. Did it end at the sky?

Newton had a hunch that gravity might extend to the moon and possibly beyond. His idea was that the moon's orbit was a kind of never-ending fall to the Earth. But unlike a falling apple, the falling moon doesn't crash to the ground because it's also simultaneously cruising sideways due to inertia. It's like one of Galileo's cannonballs, gliding sideways and falling at the same time, tracing a curved path, except that it's gliding so fast that it never reaches the surface of the spherical Earth curving away beneath it. As its orbit deviates from a straight line, the moon accelerates—not in the sense that its speed changes, but its direction of motion changes. What pulls it off a straight-line path is the incessant tug of the Earth's gravity. The resulting type of acceleration is called centripetal acceleration, a tendency to be pulled toward a center—in this case, the center of the Earth.

Newton inferred from Kepler's third law that the force of gravity weakened with distance, which explained why the more distant planets took longer to go around the sun. His calculations suggested that if the sun was pulling on the planets with the same kind of force that drew an apple to the Earth and that kept the moon in its orbit, that force had to weaken inversely with the *square* of the distance. So if the separation between the Earth and the moon could somehow be doubled, the gravitational force between them would weaken by a factor of four (two squared, not two). If the separation was tripled, the force would decrease by ninefold, not threefold. Admittedly, there were some dubious assumptions built into Newton's calculations, particularly the assumption that gravity acted instantaneously at a distance, as if the vastness of space were irrelevant. He had no idea how this could be possible, but still, the inverse-square law intrigued him.

To test it quantitatively, he estimated the centripetal acceleration of the moon as it circled the Earth at its known distance (about 60 times the radius of the Earth) and its known period of revolution (about 27 days). Then he compared the moon's acceleration to the acceleration of falling bodies on Earth, which Galileo had measured in his inclined-plane experiments. Newton found that the two accelerations differed by a factor encouragingly close to 3,600, which equals 60 squared. That was just what his inverse-square law predicted. After all, the moon was about 60 times farther from the center of the Earth than an apple falling from a tree on the Earth's surface, so its acceleration should be about 60 squared times less. In later years, Newton recalled that he'd "compared the force requisite to keep the Moon in her Orb with the force of gravity at the surface of the earth, & found them answer pretty nearly."

The notion that the tug of gravity might extend to the moon was a wild idea at the time. Remember that in Aristotelian doctrine, everything below the moon was held to be corruptible and imperfect, and everything beyond the moon was perfect, eternal, and unchanging. Newton shattered this paradigm. He unified heaven and earth and showed that the same laws of physics described both.

About twenty years after his insight with the inverse-square law, Newton took a break from his interests in alchemy and biblical chronology and revisited the question of motion due to gravity. He'd been provoked by his colleagues and rivals at the Royal Society of London. They'd challenged him to solve a much harder problem than any he'd previously considered and that none of them knew how to solve: If there was a force of attraction emanating from the sun that weakened according to an inverse-square law, how would the planets move? "In ellipses," Newton is said to have replied at once when his friend Edmond Halley posed the question. "But," asked a flabbergasted Halley, "how do you know?" "Why, I have calculated it," said Newton. When Halley urged him to explain his reasoning, Newton set about reconstructing his old work. In a furious torrent of activity, a creative outpouring almost as frenzied as what he had done as a student during the plague years, Newton wrote the *Principia*.

By assuming his laws of motion and gravity as axioms and using his calculus as a deductive instrument, Newton proved that all three of Kepler's laws followed as logical necessities. The same was true for Galileo's law of inertia, the isochronism of pendulums, the odd-number rule for balls rolling down ramps, and the parabolic arcs of projectiles. Each of them was a corollary of the inverse-square law and $F = ma$. This appeal to deductive reasoning shocked Newton's colleagues and disturbed them on philosophical grounds. Many of them were empiricists. They thought that logic applied only within mathematics itself. Nature had to be studied by experiment and observation. They were dumbfounded by the thought that nature had an inner mathematical core and that phenomena in nature could be deduced by logic from empirical axioms like the laws of gravity and motion.

The Two-Body Problem

The question that Halley posed to Newton was monstrously difficult. It required the conversion of local information into global information, the central difficulty of integral calculus and prediction that we discussed in chapter 7.

Think about what would be involved in predicting the gravitational interplay of two bodies. To simplify the problem, pretend that one of them, the sun, is infinitely massive and hence motionless, while the other, the planet in orbit, moves around it. Initially, the planet is at some distance from the sun, at a given location, and moving with a given speed in a given direction. In the next instant, the planet's velocity carries it to its next location, an infinitesimal distance from where it was a moment ago. Since it's now at a slightly different place, it feels a slightly different gravitational pull from the sun, different in both direction and magnitude. That new force (computable from the inverse-square law) tugs the planet again and changes its speed and its direction of travel by another infinitesimal amount (computable from $F = ma$) during the next infinitesimal increment of time. The process continues ad infinitum. All these infinitesimal local steps have to be integrated somehow, added together to produce the whole orbit of the moving planet.

Integrating $F = ma$ for the two-body problem is thus an exercise in the use of the Infinity Principle. Archimedes and others had applied the Infinity Principle to the mystery of curves, but Newton was the first to apply it to the mystery of motion. As hopeless as the two-body problem seemed, Newton managed to solve it with the help of the fundamental theorem of calculus. Instead of inching the planet forward instant by instant in his mind, he used calculus to thrust it forward by leaps and bounds, as if by magic. His formulas could predict where the planet would be—as well as how fast it would be moving—as far into the future as he desired.

The Infinity Principle and the fundamental theorem of calculus entered Newton's work in another novel respect. In his first attack on the two-body problem, he had idealized the planet and the sun as point-like particles. Could he model them more realistically as the colossal spherical balls that they actually were and still solve the problem? And if he could, would his results change?

This was another extraordinarily difficult calculation at that time in the development of calculus. Consider what would be needed to tally up the net tug of the giant sphere of the sun on the smaller but still giant sphere of the Earth. Every atom in the sun pulls on every atom in the Earth. The difficulty is that all those atoms are at different distances from one another. The atoms at the back of the sun are farther away, and hence exert a weaker gravitational pull on the atoms of the Earth, than the atoms in the front of the sun. Moreover, the atoms on the left and right sides of the sun pull the Earth in conflicting directions and with varying strengths depending on their own distances from the Earth. All of these effects have to be added up. Putting the pieces back together again for this problem was harder than anything anyone had ever done in integral calculus. When we solve it today, we use a method called triple integration. It's a bear.

Newton managed to solve this triple integral and found something so beautiful and so simple, it is almost unbelievable, even today. He found that he could get away with pretending that all the mass of the spherical sun was concentrated at its center; likewise for the Earth. His calculations showed that the orbit of the Earth would be the same either way. In other words, he could replace the giant spheres with infinitesimal points without incurring any error. How's that for a lie that reveals the truth!

There were many other approximations in Newton's calculations, however, whose effects were more serious and problematic. For the sake of simplicity, he'd completely ignored the gravitational pulls exerted by all the other planets. Plus he'd continued to assume that gravity acted instantaneously. He knew that both of these approximations couldn't possibly be correct, but he didn't see any way to make progress without them. He also confessed that he had no explanation for what gravity actually was or why it obeyed the mathematical description he'd given it. He knew that his critics would be suspicious of his whole program. To make his work as convincing and persuasive as possible, he couched it in the reassuring language of geometry, the gold standard of rigor and certainty as understood at that time. But it wasn't traditional Euclidean geometry; it was a peculiar, idiosyncratic admixture of classical geometry and calculus. It was calculus in geometric clothing.

Nonetheless, he did his best to give it a classical veneer. The style of the *Principia* is old-school Euclidean. Following the format of classical geometry, Newton started from axioms and postulates—his laws of motion and gravity—and treated them as unquestioned foundation stones. On them he built an edifice of lemmas, propositions, theorems, and proofs, all deduced by logic, one from the other in an unbroken chain reaching all the way back to the axioms. Just as Euclid gave the world his immortal thirteen books of the *Elements*, Newton gave the world three books of his own. Without false modesty, he called the third one *The System of the World.*

His system depicted nature as a mechanism. In the years to come, it would often be compared to a clockwork, its gears spinning, its springs stretching, all its parts moving in sequence, a wonder of cause and effect. Applying the fundamental theorem of calculus and armed with power series, ingenuity, and luck, Newton could often solve his differential equations exactly. Instead of crabbing forward instant by instant, he could leap ahead and forecast the state of his clockwork indefinitely far into the future, just as he'd done for the two-body problem of a planet orbiting the sun.

In the centuries after Newton, his system was refined by many other mathematicians, physicists, and astronomers. It was so trusted that when the motion of a planet disagreed with its predictions, astronomers assumed they were missing something important. This was how the planet Neptune was discovered in 1846. Irregularities in the orbit of Uranus suggested the presence of an unknown planet beyond it, an unseen neighbor that was perturbing Uranus gravitationally. Calculus predicted where the missing planet should be, and when astronomers looked, there it was.

Newton Meets *Hidden Figures*

By the mid-twentieth century, it seemed that physics had finally moved on from Newtonian mechanics. Quantum theory and relativity had put the old workhorse out to pasture. Yet even then it enjoyed one last hurrah, thanks to the space race between the United States and the Soviet Union.

In the early 1960s, Katherine Johnson, the African-American mathematician and heroine of *Hidden Figures*, used the two-body problem to bring astronaut John Glenn, the first American to orbit the Earth, safely back home. Johnson broke new ground in so many ways. In her analysis, the two gravitating bodies were a spacecraft and the Earth, not a planet and the sun as they had been for Newton. She used calculus to predict the position of the moving spacecraft as it orbited the Earth rotating underneath it and to calculate its trajectory for successful reentry into the atmosphere. To do that, she needed to include complications that Newton had left out, the most vital of which was that the Earth is not perfectly spherical; it bulges slightly at the equator and flattens at the poles. Getting the details right was a matter of life and death. The space capsule had to reenter the atmosphere at the right angle or it would burn up. And it had to land at the right spot in the ocean. If it splashed down too far away from the rendezvous site, Glenn might drown in his space capsule before anyone could reach him.

On February 20, 1962, Colonel John Glenn completed three orbits of our planet, and then, guided by Johnson's calculations, he reentered the atmosphere and landed safely in the North Atlantic Ocean. He was a national hero. Years later he would be elected a US senator. Few people were aware that on the day he made history, he had refused to fly his mission until Katherine Johnson herself had checked all the last-minute calculations for it. He trusted her with his life.

Katherine Johnson was a computer for the National Aeronautics and Space Administration at a time when computers were women, not machines. She was there near the start, when she helped Alan Shepard become the first American in space, and she was there near the end, when she worked on the trajectory for the first moon landing. For decades, her work was unknown to the public. Thankfully, her pioneering contributions (and her inspiring life story) have now been recognized. In 2015, at age ninety-seven, she received the Presidential Medal of Freedom from President Barack Obama. A year later, NASA named a building after her. At the dedication ceremony, the NASA official reminded the audience that "millions of people around the world watched [Alan] Shepard's flight, but what they didn't know at the time was that the calculations that got him into space and safely home were done by today's guest of honor, Katherine Johnson."

Calculus and the Enlightenment

Newton's picture of a world ruled by mathematics reverberated far beyond science. In the humanities, it served as a foil for Romantic poets like William Blake, John Keats, and William Wordsworth. At a raucous dinner party in 1817, Wordsworth and Keats, among others, agreed that Newton had destroyed the poetry of the rainbow by reducing it to its prismatic colors. They raised their glasses in a boisterous toast: "Newton's health, and confusion to mathematics."

Newton got a warmer reception in philosophy, where his ideas influenced Voltaire, David Hume, John Locke, and other Enlightenment thinkers. They were taken with the power of reason and the explanatory successes of his system, with its clockwork universe driven by causality. His empirical-deductive approach, anchored in facts and fueled by calculus, swept away the *a priori* metaphysics of earlier philosophers (I'm looking at you, Aristotle). Beyond science, it left its mark on Enlightenment conceptions of everything from determinism and liberty to natural law and human rights.

Consider, for example, Newton's sway over Thomas Jefferson —architect, inventor, farmer, third president of the United States, and author of the Declaration of Independence. There are echoes of Newton throughout the Declaration. Right from the start, the phrase "We hold these truths to be self-evident" announces the rhetorical structure. As Euclid did in the *Elements* and as Newton did in the *Principia*, Jefferson began with the axioms, the self-evident truths of his subject. Then, by force of logic, he deduced a series of inescapable propositions from those axioms, the most important of which was that the colonies had the right to sever themselves from British rule. The Declaration justifies that separation by appealing to "the Laws of Nature and of Nature's God." (Incidentally, notice the post-Newtonian deism implicit in Jefferson's ordering: God comes after the laws of nature and only in a subordinate role, as "Nature's God.") The argument is clinched by the "causes which impel [the colonists] to the separation" from the British Crown. Those causes play the role of Newtonian forces, impelling the clockwork's motion, determining the effects that must follow—in this case, the American Revolution.

If all of this seems far-fetched, keep in mind that Jefferson revered Newton. In a macabre act of devotion, he acquired a copy of Newton's death mask. And after he was no longer president, Jefferson wrote to his old friend John Adams on January 21, 1812, about the pleasures of leaving politics behind: "I have given up newspapers in exchange for Tacitus and Thucydides, for Newton and Euclid; and I find myself much the happier."

Jefferson's fascination with Newtonian principles carried over to his interest in agriculture. He wondered about the best shape for the moldboard of a plow. (A moldboard is the curved part of a plow that lifts and turns the soil cut by the plowshare.) Jefferson framed the question as one of efficiency: How should the moldboard be curved so that it would encounter the least resistance to the rising sod? The surface of the moldboard needed to be horizontal in front so that it could get under the cut soil to lift it, and it should then gradually curve to become perpendicular to the ground toward the back so that it could turn the soil and push it aside.

Jefferson asked a mathematical friend of his to address this optimization problem. In many ways, the question was reminiscent of one that Newton himself had posed in the *Principia* on the shape of a solid body of least resistance to motion through water. Guided by that theory, Jefferson had a plow fitted with a wooden moldboard of his own design.

He reported in 1798 that "an experience of five years has enabled me to say, it answers in practice to what it promises in theory." It was Newtonian calculus in the service of farming.

From Discrete to Continuous Systems

For the most part, Newton had applied calculus to one or two bodies at most—a swinging pendulum, a flying cannonball, a planet circling the sun. Solving differential equations for three or more bodies was a nightmare, as he'd learned the hard way. The problem of a mutually gravitating sun, Earth, and moon had already given him a migraine. So analyzing the whole solar system was out of the question, far beyond what even Newton could do with calculus. As he put it in one of his unpublished papers, "Unless I am much mistaken, it would exceed the force of human wit to consider so many causes of motion at the same time."

But surprisingly, going even higher, all the way up to *infinitely* many particles, made differential equations tractable again . . . as long as those particles formed a continuous medium, not a discrete set. Recall the difference: A discrete set of particles is like a collection of marbles spread out on the floor. It's discrete in the sense that you could touch one marble, move your finger through empty space, touch another one, and so on. There are gaps between the marbles. In contrast, with a continuous medium like, for instance, a guitar string, you would never have to lift your finger from the string as you traced along its length. All the particles in the guitar string hang together. Not really, of course, because a guitar string, like all other material objects, is discrete and granular at the atomic scale. But in our minds, a guitar string is more aptly regarded as a continuum. This useful fiction frees us from the chore of having to contemplate trillions and trillions of particles.

It was by addressing the mysteries of how continuous media move and change—how guitar strings vibrate to make such warm music or how heat flows from warm spots to cold spots—that calculus made its next great strides toward changing the world. But first calculus had to change itself. It needed to enlarge its concept of what differential equations were and what they could describe.

Ordinary Versus Partial Differential Equations

When Isaac Newton explained the elliptical orbits of the planets and when Katherine Johnson calculated the trajectory of John Glenn's space capsule, they were both solving a class of differential equations known as *ordinary differential equations.* The word *ordinary* is not meant to be pejorative. It's the term of art for differential equations that depend on just *one* independent variable.

For example, in Newton's equations for the two-body problem, the position of a planet was a function of time. It kept changing its location from moment to moment according to the dictates of $F = ma$. That ordinary differential equation determined how much the planet's position would change during the next infinitesimal increment of time. In this example, the planet's position is the dependent variable, since it depends on time (the independent variable). Likewise, time was the independent variable in Alan Perelson's model of HIV dynamics. He was modeling how the concentration of virus particles in the blood decreased after administration of an antiretroviral drug. The issue again was changes in time—how the viral concentration changed from moment to moment. Here, concentration played the role of the dependent variable; the independent variable was still time.

More generally, an ordinary differential equation describes how something (the position of a planet, the concentration of a virus) changes infinitesimally as the result of an infinitesimal change in something else (such as an infinitesimal increment of time). What makes such an equation "ordinary" is that there is exactly one something else, one independent variable.

Curiously, it doesn't matter how many *dependent* variables there are. As long as there is only one independent variable, the differential equation is considered ordinary. For example, it takes three numbers to pinpoint the position of a spacecraft moving in three-dimensional space. Call those numbers x, y, and z. They indicate where the spacecraft is at a given time by locating it left or right, up or down, front or back, and thus telling us how far away it is from some arbitrary reference point called the origin. As the spacecraft moves, its x, y, and z coordinates change from moment to moment. Thus, they're functions of time. To emphasize their time dependence, we could write them as $x(t)$, $y(t)$, and $z(t)$.

Ordinary differential equations are perfectly tailored to discrete systems consisting of one or more bodies. They can describe the motion of a single spaceship reentering the atmosphere, a single pendulum swinging back and forth, or a single planet as it orbits the sun. The catch is that we need to idealize each of the individual bodies as a point-like object, an infinitesimal speck with no spatial extent. Doing that allows us to think of it as existing at a point with coordinates x, y, z. The same approach works if there are many point-like particles—a swarm of tiny spaceships, a chain of pendulums connected by springs, a solar system of eight or nine planets and countless asteroids. All these systems are described by ordinary differential equations.

In the centuries after Newton, mathematicians and physicists developed many ingenious techniques for solving ordinary differential equations and thus forecasting the future of the real-world systems they describe. The mathematical techniques involved extensions of Newton's ideas about power series, Leibniz's ideas about differentials, clever transformations that could allow the fundamental theorem of calculus to be invoked, and so on. This was an enormous industry, and it continues to this day.

But not all systems are discrete—or at least, not all of them are best viewed that way, as we saw with the example of a guitar string. Consequently, not all systems can be described by ordinary differential equations. To understand why not, let's have another look at our imaginary bowl of soup cooling off on the kitchen table.

A bowl of soup is, at one level, a discrete collection of molecules, all bouncing around erratically. Yet there's no hope of seeing them, measuring them, or quantifying their motion, so nobody would ever think of using ordinary differential equations to model the cooling of a bowl of soup. There are simply too many particles to deal with, and their motion is too irregular, haphazard, and unknowable.

A much more practical way to describe what's happening is to think of the soup as a continuum. This is not really true, but it's useful. In a continuum approximation, we pretend that the soup exists at every point inside the three-dimensional volume of the soup bowl. The temperature, T, at a given point (x, y, z) depends on time, t. All of this information is captured by a function $T(x, y, z, t)$. As we will see shortly, there are differential equations for describing how this function changes in space and time. Such a differential equation is not an ordinary differential equation. It can't be, because it doesn't depend on just one independent variable. In fact, it depends on four of them: x, y, z, and t. It's a new kind of beast—a *partial differential equation,* so called because each of its independent variables plays its own "part" in causing change to occur.

Partial differential equations are much richer than ordinary differential equations. They describe continuous systems moving and changing in space and time *simultaneously* or in two or more dimensions of space. Along with a cooling bowl of soup, the saggy shape of a hammock is described by such an equation. So is the spreading of a pollutant in a lake or the flow of air over the wing of a fighter plane.

Partial differential equations are extremely difficult to handle. They make ordinary differential equations, which are already difficult, seem like child's play. Yet they are also extremely important. Our lives depend on them whenever we take to the skies.

Partial Differential Equations and the Boeing 787

Modern airplane flight is a wonder of calculus. But it wasn't always so; in a simpler time, at the dawn of aviation, the first flying machines were invented by analogy with birds and kites, by engineering savvy, and by persistent trial and error. The Wright brothers, for example, used their knowledge of bicycles to devise their three-axis system for controlling airplanes in flight and overcoming their inherent instabilities.

As aircraft became increasingly sophisticated, however, it became necessary to use more sophisticated means to design them. Wind tunnels allowed engineers to test the aerodynamic properties of their flying machines without the craft leaving the ground. Scale models, in which the designer built tiny mockups of the real planes, allowed airworthiness to be tested without building costly full-size models.

After World War II, aeronautical engineers added computers to their design arsenal. The vacuum-tube behemoths that had been used for code-breaking, artillery calculations, and weather forecasting were deployed to help create modern jet aircraft. Computers could be used to solve the complex partial differential equations that inevitably arose in the design process.

The math involved could be horrendously difficult for several reasons. For one thing, the geometry of an airplane is complicated. It's not like a sphere or a kite or a balsa-wood glider. It's a much more complex shape, with wings, fuselage, engines, tail, flaps, and landing gear. Each of these deflects the air rushing past the plane at high speed. And whenever onrushing air is deflected, it exerts a force on whatever deflected it (as anyone who has ever stuck his or her hand out the window of a car speeding down the highway knows). If an airplane wing is shaped properly, the onrushing air tends to lift it. If the plane is moving fast enough down the runway, this upward force lifts the plane off the ground and keeps it aloft. But whereas lift is a force perpendicular to the direction of oncoming airflow, another kind of force—drag—acts in a direction parallel to the flow. Drag is like friction. It resists the plane's motion and slows it down, causing its engines to work harder and burn more fuel. Calculating the size of these lift and drag forces is a brutally difficult calculus problem, far beyond the ability of any human being to solve for a realistically shaped airplane. Yet such problems must be solved. They are crucial to airplane design.

Consider the Boeing 787 Dreamliner. In 2011, Boeing—the world's largest aerospace company—rolled out its next-generation midsize jet for transporting two hundred to three hundred people on long-haul flights. The plane was touted as 60 percent quieter and 20 percent more fuel efficient than the Boeing 767, which it was designed to replace. One of its most innovative features was its use of carbon-fiber-reinforced polymers in the fuselage and wings. These space-age composite materials are lighter and stronger than aluminum, steel, and titanium, the conventional materials of choice for jet aircraft. Because they're lighter than metals, they save fuel and also make it easier for the plane to fly faster.

But perhaps the most innovative thing about the Boeing 787 was the mathematical and computational foresight that went into it, which far exceeded that in the design of any other previous plane. Calculus and computers saved Boeing an enormous amount of time —simulating a new prototype is a lot faster than building it. They also saved Boeing money—computer simulations are much cheaper to run than wind-tunnel tests, the price of which has skyrocketed in the past few decades. Douglas Ball, the chief engineer of Enabling Technology and Research at Boeing, pointed out in an interview that during the design process for the Boeing 767 in the 1980s, the company built and tested seventy-seven prototype wings. Twenty-five years later, by using supercomputers to simulate the Boeing 787's wings, they had to build and test only seven of them.

Partial differential equations entered into the process in myriad ways. For example, along with their calculations of lift and drag, Boeing's applied mathematicians used calculus to anticipate how the airplane's wings would flex when moving at six hundred miles per hour. When a wing is subjected to lift, the lift force causes the wing to flex upward and twist. One phenomenon engineers want to avoid is a dangerous effect called aeroelastic flutter, a nastier version of the fluttering of venetian blinds when a breeze blows past them. In the best case, such unwanted vibrations of the wings produce a bumpy, unpleasant ride. In the worst case, the oscillations create a positive-feedback loop: as the wings flutter, they alter the airflow over them in a way that makes them flutter even more. Aeroelastic flutter has been known to damage the wings of test aircraft and to cause structural failures and crashes (as occurred once with a Lockheed F-117 Nighthawk stealth fighter during an airshow). If a severe flutter occurred on a commercial flight, it could put hundreds of passengers at risk.

The equations that govern aeroelastic flutter are closely related to those we mentioned earlier in our discussion of facial surgery. There, the modelers channeled the spirit of Archimedes when they approximated a patient's soft tissue and skull using hundreds of thousands of gem-shaped polyhedrons and polygons. In the same spirit, Boeing's mathematicians approximated a wing with hundreds of thousands of tiny cubes, prisms, and tetrahedrons. These simpler shapes played the role of elemental building blocks. Stiffness and elastic properties were assigned to each of them, just as in the facial-surgery modeling, and then the building blocks were subjected to the relevant pushes and pulls imparted by their neighbors. The partial differential equations of elasticity theory predicted how each simple element would respond to those forces. Finally, with the help of a supercomputer, all those responses were combined and used to predict the overall vibration of the wing.

Similarly, partial differential equations were used to optimize the combustion process in the aircraft engines. This is an especially complicated problem to model. It involves the interplay of three different branches of science: chemistry (the fuel undergoes hundreds of chemical reactions at high temperature); heat flow (the heat redistributes itself within the engine as chemical energy is converted into the mechanical energy spinning the turbine blades); and fluid flow (hot gases swirl in the combustion chamber, and predicting their behavior is an exceedingly difficult problem in light of the turbulence of such gases). As before, the Boeing team used an Archimedean approach—they cut the problem into pieces, solved the problem for each piece, and put the pieces together again. It's the Infinity Principle in action, the divide-and-conquer strategy on which all of calculus rests. Here it was aided by supercomputers and a numerical method known as finite element analysis. But at the heart of it all is still calculus, embodied in differential equations.

The Ubiquity of Partial Differential Equations

The application of calculus to modern science is largely an exercise in the formulation, solution, and interpretation of partial differential equations. Maxwell's equations for electricity and magnetism are partial differential equations. So are the laws of elasticity, acoustics, heat flow, fluid flow, and gas dynamics. The list goes on: the Black-Scholes model for pricing financial options, the Hodgkin-Huxley model for the spread of electrical impulses along nerve fibers—partial differential equations all.

Even at the cutting edge of modern physics, partial differential equations still provide the mathematical infrastructure. Consider Einstein's general theory of relativity. It reimagines gravity as a manifestation of curvature in the four-dimensional fabric of space-time. The standard metaphor invites us to picture space-time as a stretchy, deformable fabric, like the surface of a trampoline. Normally the fabric is pulled taut, but it can curve under the weight of something heavy placed on it, say a massive bowling ball sitting at its center. In much the same way, a massive celestial body like the sun can curve the fabric of space-time around it. Now imagine something much smaller, say a tiny marble (which represents a planet), rolling on the trampoline's curved surface. Because the surface sags under the bowling ball's weight, it deflects the marble's trajectory. Instead of traveling in a straight line, the marble follows the contours of the curved surface and orbits around the bowling ball repeatedly. That, says Einstein, is why the planets go around the sun. They're not feeling a force; they're just following the paths of least resistance in the curved fabric of space-time.

As mind-boggling as this theory is, at its mathematical core are partial differential equations. The same is true of quantum mechanics, the theory of the microscopic realm. Its governing equation, the Schrödinger equation, is a partial differential equation too. The next chapter takes a closer look at such equations to give you a feel for what they are, where they came from, and why they matter in our everyday lives. As we'll see, partial differential equations do more than describe that bowl of soup cooling off on the kitchen table. They also explain how the microwave nuked it.

10

Making Waves

Before the early 1800s, heat was a riddle. What was it, exactly? Was it a liquid like water? It did seem to flow. But you couldn't hold it in your hands or see it. You could measure it indirectly by tracking the temperature of something hot as it cooled down, but no one knew what was going on inside the cooling object.

The secrets of heat were unraveled by a man who often felt cold. Orphaned at the age of ten, Jean Baptiste Joseph Fourier was a sickly, dyspeptic asthmatic as a teenager. As an adult, he believed heat was essential to health. He kept his room overheated and swathed himself in a heavy overcoat, even in the summer. In all aspects of his scientific life, Fourier was obsessed with heat. He originated the concept of global warming and was the first to explain how the greenhouse effect regulates the Earth's average temperature.

In 1807, Fourier used calculus to solve the riddle of heat flow. He came up with a partial differential equation that allowed him to predict how the temperature of an object, such as a red-hot iron rod, would change as it cooled. Amazingly, he found he could solve problems like this no matter how erratically the rod's temperature varied along its length at the beginning of the cooling-off process. The rod could start with hot spots here and cold spots there. No sweat—Fourier's analytical method could handle it.

Imagine a long, thin, cylindrical iron rod, heated unevenly in a blacksmith's forge so that it has patches of hot and cold scattered along its length. For simplicity, assume a perfectly insulating sleeve surrounds the rod so that heat can't escape. The only way heat can flow is to diffuse along the rod's length from hot spots to cold spots. Fourier postulated (and experiments confirmed) that the rate of change of temperature at a given point on the rod was proportional to the mismatch between the temperature at that point and the average of the temperatures of its neighbors on either side of it. And when I say *neighbors,* I really mean *neighbors*—picture two points flanking the point we're focusing on, each infinitesimally close to that point.

Under these idealized conditions, the physics of heat flow is simple. If a point is cooler than its neighbors, it heats up. If it's hotter, it cools down. The greater the mismatch, the faster the temperature evens out. If a point happens to be at precisely the average of its neighbors' temperatures, everything balances, heat doesn't flow, and the temperature of that point stays the same in the next instant.

This process of comparing a point's instantaneous temperature with that of its neighbors led Fourier to a partial differential equation that's now known as the heat equation. It involves derivatives with respect to two independent variables, one for infinitesimal changes in time (t) and one for infinitesimal changes in position (x) along the rod.

The hard part about the problem Fourier set for himself is that the hot spots and cold spots could be initially arranged higgledy-piggledy. To solve such a general problem, Fourier proposed a scheme that seemed wildly optimistic, almost foolhardy. He claimed he could replace *any* initial temperature pattern with an equivalent sum of simple sine waves.

Sine waves were his building blocks. He chose them because they made the problem easier. He knew that if the temperature started in a sine-wave pattern, it would stay in that pattern as the rod cooled off.

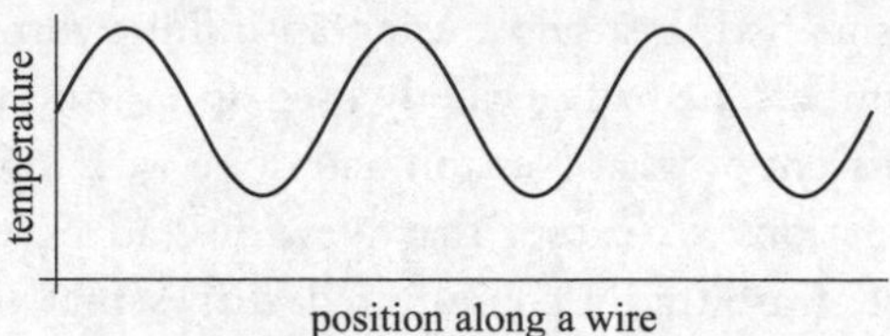

That was the key: Sine waves didn't move around. They just stood there. True, they damped down as their hot spots cooled off and their cold spots warmed up, but that decay was easy to handle. It merely meant that the temperature variations flattened out as time passed. As sketched in the diagram below, a temperature pattern that started out looking like the dashed sine wave would gradually damp down to look like the solid sine wave.

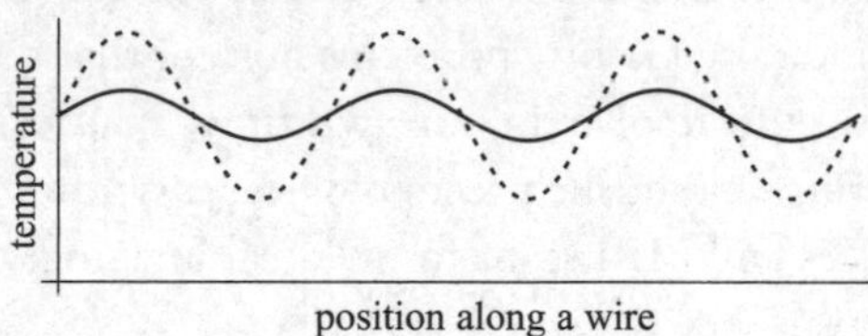

The important thing was that the sine waves stood still as they damped. They were *standing waves*.

So if he could figure out how to take an initial temperature pattern apart and break it into sine waves, he could solve the heat-flow problem for each sine wave separately. He already knew the answer to that problem: Each sine wave decayed exponentially fast at a rate that depended on how many crests and troughs it had. Sine waves with more crests decayed faster because their hot spots and cold spots were packed closer together, which made for more rapid exchange of heat between them and hence faster equilibration. Then, knowing how each sinusoidal building block decayed, all Fourier had to do was put them back together to solve the original problem.

The rub in all this was that Fourier had casually invoked an *infinite series* of sine waves. He had summoned the golem of infinity into calculus yet again, and he'd done it even more recklessly than his predecessors had. Instead of using an infinite sum of triangular shards or numbers, he had cavalierly used an infinite sum of waves. It was reminiscent of what Newton had done with his infinite sums of power functions x^n, except that Newton had never claimed he could represent arbitrarily complicated curves that included such horrors as discontinuous jumps or sharp corners in them. Fourier was now claiming exactly that—curves with corners and jumps didn't scare him. Also, Fourier's waves arose naturally from the differential equation itself, in the sense that they were its natural modes of vibration, its natural standing-wave patterns. They were tailored to heat flow. Newton's power functions had had no special claim as building blocks; Fourier's sine waves did. They were organically suited to the problem at hand.

Although his daring use of sine waves as building blocks sparked controversy and raised knotty problems of rigor that took mathematicians a century to resolve, in our own time, Fourier's big idea has played a starring role in such technologies as synthesizers for computerized voices and MRI scans for medical imaging.

String Theory

Sine waves also arise in music. They're the natural modes of vibration for the strings of guitars, violins, and pianos. A partial differential equation for such vibrations can be derived by applying Newtonian mechanics and Leibnizian differentials to an idealized model of a taut string. In this model, the string is regarded as a continuous array of infinitesimal particles stacked side by side and bonded to their neighbors by elastic forces. At any instant of time t, each particle in the string moves in accordance with the forces impinging on it. Those forces are produced by the tension in the string as neighboring particles yank on one another. Given those forces, each particle moves according to Newton's law $F = ma$. This happens at every point x along the string. Thus, the resulting differential equation depends on both x and t and is another example of a partial differential equation. It's called the wave equation because, as expected, it predicts that the typical motion of a vibrating string is a wave.

As in the heat-flow problem, certain sine waves prove useful because they regenerate themselves as they vibrate. If the ends of the string are pinned down, these sine waves don't propagate; they simply stand still and vibrate in place. If air resistance and internal friction in the string are negligible, an ideal string will vibrate forever in such a sine-wave pattern if it starts in a sine-wave pattern. And its frequency of vibration will never change. For all these reasons, sine waves continue to serve as ideal building blocks for this problem too.

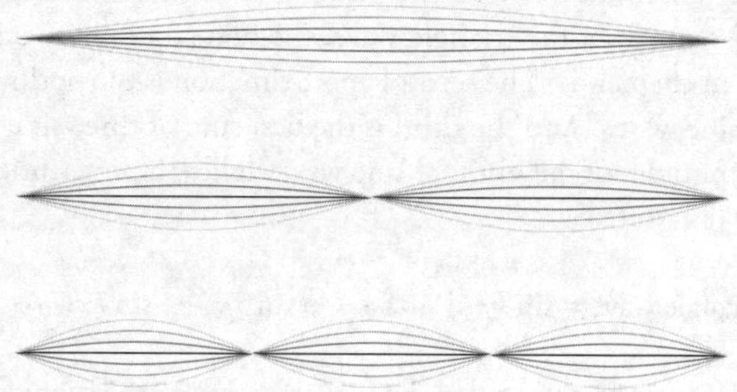

Other vibration shapes can be built out of infinite sums of sine waves. For example, in the harpsichords used in the 1700s, a string was often pulled by a quill and drawn into a triangular shape before it was released.

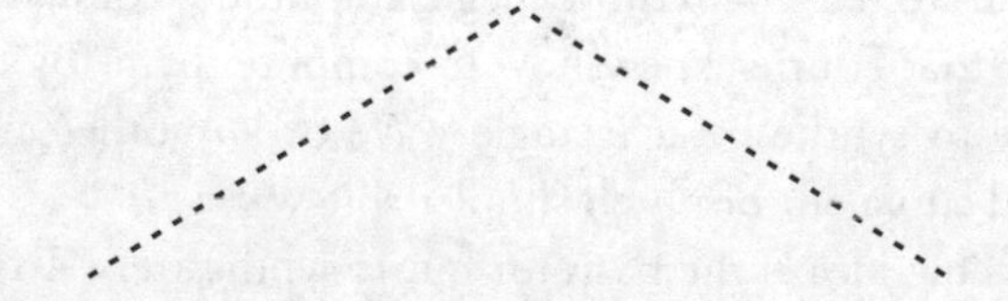

Even though a triangle wave has a sharp corner, it can be represented by an infinite sum of perfectly smooth sine waves. In other words, it doesn't take sharp corners to make sharp corners. In the diagram below, I've approximated a triangle wave, shown dashed on the bottom, with three progressively more faithful approximations by sine waves.

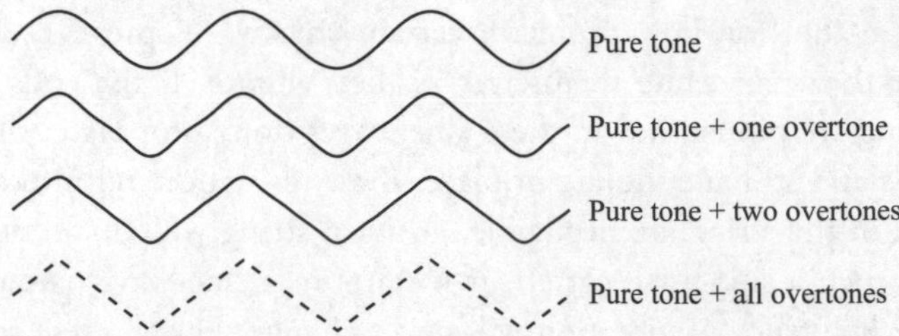

The first approximation shows a single sine wave with the best possible amplitude (*best* in the sense that it minimizes the total squared error from the triangle wave, the same optimality criterion we met in chapter 4). The second approximation is the optimal sum of two sine waves. And the third is the best sum of three sine waves. The amplitudes of the optimal sine waves follow a prescription that Fourier discovered:

$$\text{Triangle wave} = \sin x - \tfrac{1}{9}\sin 3x + \tfrac{1}{25}\sin 5x - \tfrac{1}{49}\sin 7x + \cdots.$$

This infinite sum is called the Fourier series for the triangle wave. Notice the cool numerical patterns in it. Only odd frequencies 1, 3, 5, 7, . . . appear in the sine waves, and their corresponding amplitudes are the inverse squares of the odd numbers with alternating plus and minus signs. Unfortunately, I can't easily explain why this prescription works; we would have to plow through too much nitty-gritty calculus to see where those magic amplitudes come from. But the point is that Fourier knew how to compute them. By doing so he was able to synthesize a triangle wave or any other arbitrarily complicated curve out of much simpler sine waves.

Fourier's big idea is the basis for music synthesizers. To see why, consider the sound of a note, such as the A above middle C. To generate that precise pitch, we could strike a tuning fork set to oscillate at the corresponding frequency of 440 cycles per second. A tuning fork consists of a handle and two metal tines. When the fork is hit with a rubber hammer, the tines vibrate back and forth 440 times every second. Their vibrations excite the air nearby. When a tine vibrates outward, it compresses the air; when it vibrates back, it rarefies the surrounding air. As the air molecules jiggle back and forth, they produce a sinusoidal pressure disturbance that our ears perceive as a pure tone, a boring and colorless A. It lacks what musicians call timbre. We could play the same A with a violin or a piano, and both would sound colorful and warm. Even though they too emit vibrations at a fundamental frequency of 440 cycles per second, they sound different from a tuning fork (and from each other) because of their distinct set of overtones. That's the musical term for the waves like $\sin 3x$ and $\sin 5x$ in the earlier formula for the triangle wave. Overtones add color to a note by adding in multiples of the fundamental frequency. In addition to the sine wave at 440 cycles a second, a synthesized triangle wave includes a sine-wave overtone at three times that frequency ($3 \times 440 = 1320$ cycles per second). That overtone is not as strong as the fundamental $\sin x$ mode. Its relative amplitude is only 1⁄9 as large as the fundamental, and the other odd-numbered modes are even weaker. In musical terms, these amplitudes determine the loudness of the overtones. The richness of the sound of a violin has to do with its particular combination of softer and louder overtones.

The unifying power of Fourier's idea is that the sound of *any* musical instrument can be synthesized by an array of infinitely many tuning forks. All we need to do is strike the tuning forks with the right strengths and at the right times and, incredibly, out pops the sound of a violin or a piano or even a trumpet or an oboe, although we're using nothing more than colorless sine waves. This is essentially how the first electronic synthesizers worked: they reproduced the sound of any instrument by combining a large number of sine waves.

Back in high school I took a class in electronic music that gave me a feeling for what sine waves could do. This was in the dark ages of the 1970s, when electronic music was produced by a big box that looked like an old-fashioned switchboard. My classmates and I would plug cables into various jacks and turn knobs up and down, and out would come the sound of sine waves, square waves, and triangle waves. My recollection is that sine waves had a clear, open sound, like flutes. Square waves sounded piercing, like fire alarms. Triangle waves were brassy. With one knob, we could change a wave's frequency to raise or lower its pitch. With another, we could change its amplitude to make it sound louder or softer. By plugging in several cables at once, we could add waves and their overtones together in different combinations, just as Fourier had done abstractly, but for us the experience was sensory. We could see the waves' shapes on an oscilloscope at the same time as we listened to them. You could try all this for yourself now on the internet. Search for something like *the sound of triangle waves* and you'll find interactive demos that will let you feel like you're sitting right there in my classroom in 1974, playing with waves for the fun of it.

The larger significance of Fourier's work is that he took the first step toward using calculus as a soothsayer to predict how a continuum of particles could move and change. This was an enormous advance beyond Newton's work on the motion of discrete sets of particles. In the centuries to come, scientists would extend Fourier's methods to forecast the behavior of other continuous media, like the flutter of a Boeing 787 wing, the appearance of a patient after facial surgery, the flow of blood through an artery, or the rumbling of the ground after an earthquake. Today these techniques are ubiquitous in science and engineering. They are used to analyze shock waves from a thermonuclear blast; radio waves for communications; the waves of digestion in the intestine that allow nutrients to be absorbed and send waste products moving in the right direction; the pathological electrical waves in the brain associated with epilepsy and Parkinson's tremors; and the congestion waves of traffic on a highway, as seen in the exasperating phenomenon of phantom jams, where traffic slows down for no apparent reason. Fourier's ideas and their offshoots have enabled all of these wave phenomena to be understood mathematically, sometimes with the help of formulas, other times through massive computer simulations, so we can explain, predict, and, in some cases, control or abolish them.

Why Sine Waves?

Before we leave sine waves and move on to their two- and three-dimensional counterparts, it's worth clarifying what makes them so special. After all, other types of curves can serve as building blocks, and sometimes they work better than sine waves. For instance, to capture localized features like fingerprint ridges, wavelets got the nod from the FBI. Wavelets are often superior to sine waves for many image- and signal-processing tasks in fields like earthquake analysis, art restoration and authentication, and facial recognition.

So why are sine waves so well suited to the solution of the wave equation and the heat equation and other partial differential equations? Their virtue is that they play very nicely with derivatives. Specifically, the derivative of a sine wave is another sine wave, shifted by a quarter cycle. That's a remarkable property. It's not true of other kinds of waves. Typically, when we take the derivative of a curve of any kind, that curve will become distorted by being differentiated. It won't have the same shape before and after. Being differentiated is a traumatic experience for most curves. But not for a sine wave. After its derivative is taken, it dusts itself off and appears unfazed, as sinusoidal as ever. The only injury it suffers—and it isn't even an injury, really—is that the sine wave shifts in time. It peaks a quarter of a cycle earlier than it used to.

We saw an imperfect version of this in chapter 6 when we looked at the day-to-day variations in day length in New York City in the year 2018 and compared them to the daily changes in day length, the number of additional minutes of sunlight from one day to the next. We saw that both curves looked approximately sinusoidal, except that the difference in daylight from one day to the next formed a wave that was shifted three months earlier than the data from which it came. Put simply, the longest day in 2018 was June 21, while the fastest-lengthening day was three months earlier, March 20. This is what we expect from sinusoidal data. If day-length data were a *perfect* sine wave, and if we looked at its difference not from one day to the next but from one *instant* to the next, then its instantaneous rate of change (the "derivative" wave derived from it) would itself be a perfect sine wave, shifted exactly a quarter of a cycle earlier. Back in chapter 6 we also saw why the quarter-cycle shift occurs. It follows from the deep connection between sine waves and uniform circular motion. (You may want to look back at that argument if it seems hazy now.)

That quarter-cycle shift has a fascinating consequence. It implies that if we take *two* derivatives of the sine wave, it shifts a quarter plus another quarter cycle earlier. So in total it gets shifted *half* a cycle earlier. That means that its former peak is now a valley, and vice versa. The sine wave has turned upside down. In mathematical terms, this is expressed by the formula

$$\frac{d}{dx}\left(\frac{d}{dx}\sin x\right) = -\sin x$$

where the Leibnizian differentiation symbol d/dx means "take the derivative of whatever expression appears to the right." The formula shows that taking two derivatives of $\sin x$ amounts to nothing more than multiplying it by -1. This replacement of two derivatives by a simple multiplication is a fantastic simplification. Taking two derivatives is a full-bore calculus operation, whereas multiplying by -1 is middle-school arithmetic.

But why, you may be asking yourself, would anyone ever want to take two derivatives of something? Because nature does—and it does it all the time. Or, rather, our *models* of nature do it all the time. For example, in Newton's law of motion, $F = ma$, the acceleration a involves two derivatives. To see why, remember that the acceleration is the derivative of speed and speed is the derivative of distance. That makes acceleration the *derivative of the derivative* of distance, or to put it more concisely, the second derivative of distance. Second derivatives come up everywhere in physics and engineering. Along with Newton's equation, they also star in the heat equation and the wave equation.

So *that's* why sine waves are so well suited to those equations. For sine waves, two derivatives boil down to mere multiplication by -1. In effect, the inherent calculus that made the heat and wave equations hard to analyze is no longer an issue when we restrict our attention to sine waves. The calculus gets stripped out and replaced by multiplication. This is what made the vibrating-string problem and the heat-flow problem so much easier to solve for sine waves. If an arbitrary curve could be built from them, that curve would inherit the virtues of sine waves. The only hitch was that infinitely many sine waves needed to be added together to build up an arbitrary curve, but that was a small price to pay.

This is the calculus perspective on why sine waves are special. Physicists have their own perspective, one that is also worth understanding. To a physicist, what's remarkable about sine waves (in the context of the vibration and heat flow problems) is that they form *standing waves*. They don't travel along the string or the rod. They remain in place. They oscillate up and down but never propagate. Even more remarkably, standing waves vibrate at a unique frequency. That's a rarity in the world of waves. Most waves are a combination of many frequencies, just as white light is a combination of all the colors of the rainbow. In that respect, a standing wave is pure, not a mixture.

Visualizing Modes of Vibration: Chladni Patterns

The warm sound of a guitar and the plaintive sound of a violin are related to the vibrations set up in the belly and body of the instrument, in the wood and the cavities inside, where sound waves vibrate and resonate. Those vibration patterns determine the quality and voice of the instrument. That's part of what makes a Stradivarius so special, its uniquely evocative vibration patterns of wood and air. We still don't understand exactly what makes certain violins sound better than others, but the key must be something about their modes of vibration.

In 1787, a German physicist and musical-instrument maker named Ernst Chladni published an article showing a clever way to visualize these vibrational patterns. Instead of using a shape as complicated as a guitar or a violin, though, he played a much simpler instrument—a thin metal plate—by drawing a violin bow across its edge. In this way, he was able to get the plate to vibrate and sing (a bit like the way you can get a half-filled wineglass to sing by rubbing your finger around its rim). To visualize the vibrations, Chladni sprinkled a fine dust of sand onto the plate before he bowed it. When he stroked the plate, the sand bounced off the parts that were vibrating the most and settled in the parts that weren't vibrating at all. The resulting curves are now called Chladni patterns.

You may have seen a demonstration of Chladni patterns at science museums. A metal plate is placed over a loudspeaker and covered with sand, then it's driven to vibrate by an electronic signal generator. As the frequency of the sound coming out of the loudspeaker is adjusted, the plate can be excited into different resonant patterns. Whenever the loudspeaker tunes into a new resonant frequency, the sand rearranges itself into a different standing-wave pattern. The plate divides itself into neighboring regions that vibrate in opposite directions, bounded by nodal curves where the plate remains motionless.

Perhaps it seems odd that some parts of the plate don't move. But that shouldn't be surprising. We saw the same thing with sine waves on a string. The points where the string doesn't move are the nodes of vibration. For a plate, there are similar nodes, except they are not isolated points. Rather, they link together to form nodal lines and curves. These are the curves that Chladni made manifest in his experiments. They were considered so astonishing at the time that Chladni was invited to show them to Emperor Napoleon himself. Napoleon, who had some training in math and engineering, was so intrigued that he established a contest and challenged the greatest mathematicians of Europe to explain Chladni's patterns.

The necessary mathematics did not exist at that time. The preeminent mathematician in Europe, Joseph Louis Lagrange, felt that the problem was beyond reach and that no one would solve it. Indeed, only one person tried. Her name was Sophie Germain.

The Noblest Courage

Sophie Germain had taught herself calculus at a young age. The daughter of a wealthy family, she had become entranced by mathematics after reading a book about Archimedes in her father's library. When her parents found out that she loved mathematics and was staying up late at night to work on it, they took away her candles, left her fire unlit, and confiscated her nightgowns. Sophie persisted. She wrapped herself in quilts and worked by the light of stolen candles. Eventually her family relented and gave her their blessing.

Germain, like all women of her era, was not permitted to attend university, so she continued to teach herself, in some cases by obtaining lecture notes from the courses at the nearby École Polytechnique using the name Monsieur Antoine-August Le Blanc, a student who had left the school. Unaware of his departure, academy administrators continued to print lecture notes and problem sets for him. She submitted work under his name until one of the school's teachers, the great Lagrange, noticed the remarkable improvement in Monsieur Le Blanc's previously abysmal performance. Lagrange requested a meeting with Le Blanc and was delighted and astonished to discover her true identity. He took Germain under his wing.

Her earliest triumphs were in number theory, where she made important contributions to one of the most difficult unsolved problems in that field, known as Fermat's last theorem. When she felt she'd made a breakthrough, she wrote to the world's greatest number theorist (and one of the greatest mathematicians of all time), Carl Friedrich Gauss, once more using the pseudonym of Antoine Le Blanc. Gauss admired the brilliance of his mysterious correspondent and they conducted a lively exchange of letters for three years. Matters darkened one day in 1806 when events threatened Gauss's life. Napoleon's army had begun storming through Prussia, and Gauss's home city of Brunswick was taken. Using family connections, Germain wrote to a friend who was a general in the French army and asked him to guarantee Gauss's safety. When word got back to Gauss that his life had been protected by the intervention of a Mademoiselle Sophie Germain, he was grateful but puzzled, since he knew no one by that name. In her next letter, Germain unmasked herself. Gauss was flabbergasted to learn that he had been corresponding with a woman. Given the depth of her insights and recognizing all the prejudices and obstacles she must have endured, he told her that "without doubt she must have the noblest courage, quite extraordinary talents and superior genius."

So when she heard of the competition to solve the mystery of Chladni patterns, Germain rose to the challenge. She was the only person brave enough to take a stab at developing the necessary theory from scratch. Her solution involved creating a new subfield of mechanics, the theory of elasticity for flat, thin, two-dimensional plates, going beyond the earlier and much simpler theories for one-dimensional strings and beams. She built it on principles of forces and displacements and curvatures, and she used techniques of calculus to formulate and solve the relevant partial differential equations for Chladni's vibrating plates and the marvelous patterns they produced. But given the gaps in Germain's education and her lack of formal training, her attempted solution contained flaws that the judges noticed. They felt that the problem had not been fully solved, and they renewed the contest for another two years, and then another two after that. On her third try, Germain was awarded the prize, the first woman ever to be so honored by the Paris Academy of Sciences.

Microwave Ovens

Chladni patterns allow us to visualize standing waves in two dimensions. In our daily lives, we rely on the three-dimensional counterpart of Chladni patterns whenever we use a microwave oven. The inside of a microwave oven is a three-dimensional space. When you press the start button, the oven fills with a standing-wave pattern of microwaves. Though you can't see these electromagnetic vibrations with your eyes, you can visualize them indirectly by mimicking what Chladni did with his sand.

Take a microwave-safe plate and cover it completely with a thin layer of processed shredded cheese (or anything else that will lie flat and melt easily, like a thin slab of chocolate or a sprinkling of mini-marshmallows). Before you put the plate in the oven, be sure to take out the rotating turntable. That's important because you want the plate of cheese (or whatever you're using) to stand still to allow you to detect the hot spots. Once the turntable is out and the plate is inside, close the door and turn on the microwave. Let it go for about thirty seconds, no more. Then take out the plate. You'll see places where the cheese has melted completely. Those are the hot spots. They correspond to anti-nodes of the microwave pattern, the places where the vibrations are most vigorous. They're like the peaks and troughs of a sine wave or like the places in the Chladni pattern where the sand is *not* (because the vigorous oscillations have shaken it off).

For a standard microwave oven that runs at 2.45 GHz (meaning the waves vibrate back and forth 2.45 billion times a second), you should find that the distance between neighboring melted spots is about two and a half inches, or six centimeters. Keep in mind, that's only the distance from a peak to a trough and hence is *half* a wavelength. To get the full wavelength, we double that distance. Thus the wavelength of the standing-wave pattern in the oven is about five inches, or twelve centimeters.

Incidentally, you can use your oven to calculate the speed of light. Multiply the frequency of vibration (listed on the oven's door frame) by the wavelength you measured in your experiment, and you should get the speed of light or something pretty close to it. Here's how it would go with the numbers I just gave: The frequency is 2.45 billion cycles per second. The wavelength is 12 centimeters (per cycle). Multiplying them together gives 29.4 billion centimeters per second. That's pretty close to the accepted value for the speed of light, 30 billion centimeters per second. Not bad for such a crude measurement.

Why Microwave Ovens Used to Be Called Radar Ranges

At the end of World War II, the Raytheon Company was looking for new applications for its magnetrons, the high-powered vacuum tubes used in radar. A magnetron is the electronic analog of a whistle. Just as a whistle sends out sound waves, a magnetron sends out electromagnetic waves. These waves can be bounced off an airplane overhead to detect how far away it is and how fast it's moving. Nowadays, radar is used to track the movement of everything from ships and speeding cars to fastballs, tennis serves, and weather patterns.

After the war, in 1946, Raytheon had no idea what it was going to do with all the magnetrons it had been manufacturing. An engineer named Percy Spencer noticed one day that a peanut-cluster bar in his pocket had turned into a gooey, sticky mess while he was working with a magnetron. He realized that the microwaves it emitted could warm food very effectively. To explore the idea further, he tried pointing a magnetron at an egg, and it got so hot it exploded in someone's face. Spencer also demonstrated that he could make popcorn with it. This connection between radar and microwaves is why the first microwave ovens were called radar ranges. They were not a commercial hit until the late 1960s. The first microwave ovens were too big, almost six feet tall, and extremely expensive, costing the equivalent of tens of thousands of dollars in today's money. But eventually microwaves became sufficiently miniaturized and cheap enough that ordinary families could afford them. Today, at least 90 percent of households in industrialized countries have them.

The story of radar and microwave ovens is a testament to the interconnectedness of science. Think of what went into them: physics, electrical engineering, materials science, chemistry, and good old serendipitous invention. Calculus played an important part too. It provided the language for describing waves and the tools for analyzing them. The discovery of the wave equation, which started as an outgrowth of music in connection with vibrating strings, was ultimately used by Maxwell to predict the existence of electromagnetic waves. From there it was a short hop to vacuum tubes, transistors, computers, radar, and microwave ovens. Along the way, Fourier's methods were indispensable. And as we are about to see, his techniques played a role in the discovery of a new use for higher-energy electromagnetic waves. These much more energetic waves were discovered by accident at the turn of the twentieth century. No one was sure what they were, so in honor of the unknown in mathematics, they were named x-rays.

Computerized Tomography and Brain Imaging

Microwaves are good for cooking, but x-rays are good for seeing into our bodies. They allow noninvasive diagnosis of broken bones, skull fractures, and curved spines. Unfortunately, traditional x-rays captured on black-and-white film are insensitive to subtle variations in tissue density. This limits their usefulness for examining soft tissues and organs. A more modern form of medical imaging, called CT scanning, is hundreds of times more sensitive than conventional x-ray films. Their precision has revolutionized medicine.

The *C* stands for *computerized* and the *T* stands for *tomography*, meaning the process of visualizing something by cutting it into slices. A CT scan uses x-rays to image an organ or a tissue one slice at a time. When a patient is placed in a CT scanner, x-rays are sent through the person's body at many different angles and recorded by a detector on the other side. From all that information—from all those views at different angles—it's possible to reconstruct much more clearly what the x-rays passed through. In other words, CT is not just a matter of seeing; it's a matter of inferring, deducing, and calculating. Indeed, the most brilliant and revolutionary part of CT is its use of sophisticated mathematics. With the help of calculus, Fourier analysis, signal processing, and computers, the CT software infers the properties of the tissue, organ, or bone through which the x-rays passed and then generates a detailed picture of that part of the body.

To see how calculus plays a role in all this, first we need to understand what problem CT solves and how it solves it.

Imagine firing a beam of x-rays through a slice of brain tissue. As the x-rays travel, they encounter gray matter, white matter, possibly brain tumors, blood clots, and so on. These tissues absorb the x-rays' energy to a greater or lesser degree, depending on the type of tissue it is. The goal of CT is to map the absorption pattern in the whole slice. From that information, CT can reveal where tumors or clots may be. CT doesn't see the brain directly; it sees the x-ray–absorption pattern in the brain.

The math works like this. As an x-ray travels through a given point in the brain slice, it loses some of its intensity. This loss of intensity is like what happens when ordinary light passes through sunglasses and becomes less bright. The complication here is that there is a sequence of different brain tissues along the x-ray's path, so the tissues act more like a sequence of sunglasses, one in front of another, all of different opacities. And we don't know the opacity of any of the sunglasses; that's what we're trying to figure out!

Because of this variability in the absorption properties of the different tissues, when the x-rays emerge from the brain and strike the x-ray detector on the other side, their intensity has been reduced by disparate amounts along the way. To compute the net effect of all of these reductions, we have to figure out how much the intensity was reduced, step by infinitesimal step as the x-rays traveled through the tissue, and then combine all the results appropriately. This computation amounts to an integral.

The appearance of integral calculus here shouldn't come as a surprise. It's the most natural way to make this very complicated problem more tractable. As always, we appeal to the Infinity Principle. First, we imagine chopping the x-rays' path into infinitely many infinitesimal steps, then we figure out how much their intensity attenuates with each step, and finally we put all the answers back together to compute the net attenuation along the given line of travel.

Sadly, having done this, we've obtained only a single piece of information. We know the total attenuation of the x-rays only along the particular path that the x-rays followed. That doesn't tell us much about the brain slice as a whole. It doesn't even tell us much about the particular line the x-rays traveled on. It just tells us the net attenuation along the line, not the point-to-point pattern of attenuation along it.

Let me try to illustrate the difficulty by analogy: Think about all the different ways we could add up numbers to make 6. Just as the number 6 can result from 1 + 5 or 2 + 4 or 3 + 3, the same net attenuation of the x-rays could result from many different sequences of local attenuations. For example, there could be high attenuation at the beginning of the line and low attenuation at the end. Or it could be the other way around. Or there could be a constant, medium level of attenuation the whole way through. We have no way of distinguishing among these possibilities from just one measurement.

However, once we recognize the difficulty, we can immediately see how to solve it. We need to fire x-rays along *many* different directions. That's the heart of computerized tomography. By firing x-rays from multiple directions through the same point of tissue and then repeating the measurement for many different points, we should, in principle, be able to map out the attenuation factors everywhere in the brain. This is not quite the same thing as *looking* at the brain, but it's almost as good, because it provides information about which types of tissues occur in which brain regions.

The mathematical challenge, then, is to reassemble the information obtained from all the measurements along lines into a coherent two-dimensional picture of the whole brain slice. This is where Fourier analysis came in. It allowed a South African physicist named Allan Cormack to solve the reassembly problem. Fourier analysis entered because there was a circle lurking in the problem. That circle was the circle of all the lines—all the directions along which the x-rays could be fired, edge on, into a two-dimensional slice.

Remember that circles are always associated with sine waves, and sine waves are the building blocks of Fourier series. By writing the reassembly problem in terms of Fourier series, Cormack was able to boil a two-dimensional reassembly problem down to an easier one-dimensional problem. In effect, he got rid of the 360 degrees of possible angles. Then, with great prowess in integral calculus, he managed to solve the one-dimensional reassembly problem. The upshot was that, given the measurements along a full circle of lines, he could deduce the properties of the tissue inside. He could infer the absorption map. It was almost like seeing the brain itself.

In 1979, Cormack shared the Nobel Prize in Physiology or Medicine with Godfrey Hounsfield for their development of computer-assisted tomography. Neither of them was a medical doctor. Cormack developed the Fourier-based mathematical theory of CT scanning in the late 1950s. Hounsfield, a British electrical engineer, invented the scanner in collaboration with radiologists in the early 1970s.

The invention of the scanner provides another demonstration of the unreasonable effectiveness of mathematics. In this case, the ideas that made CT scanning possible had existed for more than half a century and had no connection whatsoever to medicine.

The next part of the story began in the late 1960s. Hounsfield had already tested a prototype of his invention on pigs' brains. He was desperate to find a clinical radiologist to help him extend his work to human patients, but one doctor after another refused to meet with him. They all thought he was a crackpot. They knew soft tissues couldn't be visualized with x-rays. A traditional x-ray of a head, for example, showed the skull bones clearly, but the brain looked like a featureless cloud. Tumors, hemorrhages, and blood clots weren't visible, despite what Hounsfield claimed.

Finally, one radiologist agreed to hear him out. The conversation didn't go well. At the end of the meeting, the skeptical radiologist handed Hounsfield a jar containing a human brain with a tumor in it and challenged him to image it with his scanner. Hounsfield soon brought back images of the brain that pinpointed not only the tumor but also areas of bleeding within it.

The radiologist was stunned. Word spread, and soon other radiologists came on board. When Hounsfield published the first computerized tomographs in 1972, they shocked the medical world. Radiologists could suddenly use x-rays to see tumors, cysts, gray matter, white matter, and the fluid-filled cavities of the brain.

Ironically, given that wave theory and Fourier analysis began with the study of music, at a key moment in the development of computerized tomography, music proved indispensable again. Hounsfield had his breakthrough ideas in the mid-1960s when he was working for a company called Electric and Musical Industries. He had first worked on EMI's radar and guided weaponry, and then he turned his attention to developing Britain's first all-transistor computer. After that smashing success, EMI decided to support Hounsfield and let him do whatever he wanted for his next project. At that time, EMI was flush with money and could afford to take risks. Their profits had doubled after they'd signed a band from Liverpool called the Beatles.

Hounsfield approached management with his idea of imaging organs with x-rays, and EMI's deep pockets helped him take the first step. He came up with his own approach to solving the reassembly problem in the mathematics, unaware that Cormack had solved it a decade earlier. And Cormack, in turn, didn't know that a pure mathematician named Johann Radon had solved it forty years before him, with no application in mind. The quest for pure mathematical understanding had given CT scanning the tools it needed, half a century ahead of time.

In Cormack's Nobel Prize address, he mentioned that he and his colleague Todd Quinto had looked into Radon's results and were trying to generalize them to three- and even four-dimensional regions. That must have been hard for his audience to fathom. We live in a three-dimensional world. Why would anyone want to study a four-dimensional brain? Cormack explained:

> What is the use of these results? The answer is that I don't know. They will almost certainly produce some theorems in the theory of partial differential equations, and some of them may find application in imaging with MRI or ultrasound, but that is by no means certain. It is also beside the point. Quinto and I are studying these topics because they are interesting in their own right as mathematical problems, and that is what science is all about.

11

The Future of Calculus

THE TITLE OF this chapter might raise a few eyebrows among those who believe that calculus is finished. How could it have a future? It's over now, isn't it? This is something you hear surprisingly often in mathematical circles. According to this narrative, calculus began with a bang, thanks to the breakthroughs of Newton and Leibniz. Their discoveries sparked a gold-rush mentality in the 1700s, a period marked by playful, almost giddy exploration during which the golem of infinity was allowed to run wild. By giving it free rein, mathematicians produced a raft of spectacular results but also generated a lot of nonsense and confusion. So in the 1800s, the next few generations of mathematicians, a more rigorous lot, prodded the golem back into its cage. They expunged infinity and infinitesimals from calculus, shored up the foundations of the subject, and finally clarified what limits, derivatives, integrals, and real numbers actually meant. By around 1900, their mopping-up operation was complete.

To my mind, that vision of calculus is far too blinkered. Calculus is not just the work of Newton and Leibniz and their successors. It started much earlier than that and it's still going strong today. Calculus, to me, is defined by its credo: to solve a hard problem about anything continuous, slice it into infinitely many parts and solve them. By putting the answers back together, you can make sense of the original whole. I've called this credo the Infinity Principle.

The Infinity Principle was there from the beginning, in

Archimedes's work on curved shapes, and it was there in the scientific revolution, in Newton's system of the world, and it's with us today in our homes, at our jobs, and in our cars. It helped give us GPS, cell phones, lasers, and microwave ovens. The FBI used it to compress millions of fingerprint files. Allan Cormack used it to create the theory for CT scanning. Both the FBI and Cormack solved a hard problem by reassembling it from simpler parts: wavelets for fingerprints, sine waves for CT. From this point of view, calculus is the sprawling collection of ideas and methods used to study anything—any pattern, any curve, any motion, any natural process, system, or phenomenon—that changes smoothly and continuously and hence is grist for the Infinity Principle. This broad definition goes far beyond the calculus of Newton and Leibniz to include its descendants: multivariable calculus, ordinary differential equations, partial differential equations, Fourier analysis, complex analysis, and any other part of higher mathematics where limits, derivatives, and integrals appear. Viewed this way, calculus is not over. It's as hungry as ever.

But I'm in the minority here. Actually, a minority of one. None of my colleagues in the math department would agree that all of this is calculus, and for good reason: It would be absurd. Half the courses in the curriculum would have to be renamed. Along with Calculus 1, 2, and 3, we'd now have Calculus 4 through 38. Not very descriptive. So instead, we give different names to each offshoot of calculus and obscure the continuity among them. We slice the whole of calculus into its smallest consumable parts. That's ironic, or perhaps fitting, given that calculus itself is about slicing continuous things into parts to make them easier to understand. Let me be clear: I have no objection to all the different course names. All I'm saying is that slicing can be misleading when it makes us forget that the parts belong together, that they're all part of something bigger. My goal in this book has been to show calculus as a whole, to give a feeling for its beauty, unity, and grandeur.

What, then, might the future hold for calculus? As they say, prediction is always difficult, especially about the future, but I think it's safe to assume that several trends are likely to be important in the years ahead. These include

- New applications of calculus to the social sciences, music, the arts, and the humanities
- Ongoing applications of calculus to medicine and biology
- Coping with the randomness inherent in finance, economics, and the weather
- Calculus in the service of big data, and vice versa
- The continuing challenge of nonlinearity, chaos, and complex systems
- The evolving partnership between calculus and computers, including artificial intelligence
- Pushing the boundaries of calculus in the quantum realm.

This is a lot of ground to cover. Rather than saying a little about each of the topics mentioned here, I'll focus on a few of them. After a brief foray into the differential geometry of DNA, where the mystery of curves meets the secret of life, we'll consider some case studies that I hope you'll find philosophically provocative. These include the challenges to insight and prediction caused by the rise of chaos, complexity theory, computers, and artificial intelligence. For all of that to make sense, however, we will need to review the fundamentals of nonlinear dynamics. Examining that context will allow us to better appreciate the challenges ahead.

The Writhing Number of DNA

Calculus has traditionally been applied in the "hard" sciences like physics, astronomy, and chemistry. But in recent decades, it has made inroads into biology and medicine, in fields like epidemiology, population biology, neuroscience, and medical imaging. We've seen examples of mathematical biology throughout our story, ranging from the use of calculus in predicting the outcome of facial surgery to the modeling of HIV as it battles the immune system. But all those examples were concerned with some aspect of the mystery of change, the most modern obsession of calculus. In contrast, the following example is drawn from the ancient mystery of curves, which was given new life by a puzzle about the three-dimensional path of DNA.

The puzzle had to do with how DNA, an enormously long molecule that contains all the genetic information needed to make a person, is packaged in cells. Every one of your ten trillion or so cells contains about two meters of DNA. If laid end to end, that DNA would reach to the sun and back dozens of times. Still, a skeptic might argue that this comparison is not as impressive as it sounds; it merely reflects how many cells each of us has. A more informative comparison is with the size of the cell's nucleus, the container that holds the DNA. The diameter of a typical nucleus is about five-millionths of a meter, and it is therefore four hundred thousand times smaller than the DNA that has to fit inside it. That compression factor is equivalent to stuffing twenty miles of string into a tennis ball.

On top of that, the DNA can't be stuffed into the nucleus haphazardly. It mustn't get tangled. The packaging has to be done in an orderly fashion so the DNA can be read by enzymes and translated into the proteins needed for the maintenance of the cell. Orderly packaging is also important so that the DNA can be copied neatly when the cell is about to divide.

Evolution solved the packaging problem with spools, the same solution we use when we need to store a long piece of thread. The DNA in cells is wound around molecular spools made of specialized proteins called histones. To achieve further compaction, the spools are linked end to end, like beads on a necklace, and then the necklace is coiled into ropelike fibers that are themselves coiled into chromosomes. These coils of coils of coils compact the DNA enough to fit it into the cramped quarters of the nucleus.

But spools were not nature's original solution to the packaging problem. The earliest creatures on Earth were single-celled organisms that lacked nuclei and chromosomes. They had no spools, just as today's bacteria and viruses don't. In such cases, the genetic material is compacted by a mechanism based on geometry and elasticity.

Imagine pulling a rubber band tight and then twisting it from one end while holding it between your fingers. At first, each successive turn of the rubber band introduces a twist. The twists accumulate, and the rubber band remains straight until the accumulated torsion crosses a threshold. Then the rubber band suddenly buckles into the third dimension. It begins to coil on itself, as if writhing in pain. These contortions cause the rubber band to bunch up and compact itself. DNA does the same thing.

This phenomenon is known as supercoiling. It is prevalent in circular loops of DNA. Although we tend to picture DNA as a straight helix with free ends, in many circumstances it closes on itself to form a circle. When this happens, it's like taking off your belt, putting a few twists in it, and then buckling it closed again. After that the number of twists in the belt cannot change. It is locked in. If you try to twist the belt somewhere along its length without taking it off, countertwists will form elsewhere to compensate. There is a conservation law at work here. The same thing happens when you store a garden hose by piling it on the floor with many coils stacked on top of each other. When you try to pull the hose out straight, it twists in your hands. Coils convert to twists. The conversion can also go in the other direction, from twists to coils, as when a rubber band writhes when twisted. The DNA of primitive organisms makes use of this writhing. Certain enzymes can cut DNA, twist it, and then close it back up. When the DNA relaxes its twists to lower its energy, the conservation law forces it to become more supercoiled and therefore more compact. The resulting path of the DNA molecule no longer lies in a plane. It writhes about in three dimensions.

In the early 1970s an American mathematician named Brock Fuller gave the first mathematical description of this three-dimensional contortion of DNA. He invented a quantity that he dubbed the *writhing number* of DNA. He derived formulas for it using integrals and derivatives and proved certain theorems about the writhing number that formalized the conservation law for twists and coils. The study of the geometry and topology of DNA has been a thriving industry ever since. Mathematicians have used knot theory and tangle calculus to elucidate the mechanisms of certain enzymes that can twist DNA or cut it or introduce knots and links into it. These enzymes alter the topology of DNA and hence are known as topoisomerases. They can break strands of DNA and reseal them, and they are essential for cells to divide and grow. They have proved to be effective targets for cancer-chemotherapy drugs. The mechanism of action is not completely clear, but it is thought that by blocking the action of topoisomerases, the drugs (known as topoisomerase inhibitors) can selectively damage the DNA of cancer cells, which causes them to commit cellular suicide. Good news for the patient, bad news for the tumor.

In the application of calculus to supercoiled DNA, the double helix is modeled as a continuous curve. As usual, calculus likes to work with continuous objects. In reality, DNA is a discrete collection of atoms. There's nothing truly continuous about it. But to a good approximation, it can be treated as if it were a continuous curve, like an ideal rubber band. The advantage of doing that is that the apparatus of elasticity theory and differential geometry, two spinoffs of calculus, can then be applied to calculate how DNA deforms when subjected to forces from proteins, from the environment, and from interactions with itself.

The larger point is that calculus is taking its usual creative license, treating discrete objects as if they were continuous to shed light on how they behave. The modeling is approximate but useful. Anyway, it's the only game in town. Without the assumption of continuity, the Infinity Principle cannot be deployed. And without the Infinity Principle, we have no calculus, no differential geometry, and no elasticity theory.

I expect in the future we will see many more examples of calculus and continuous mathematics being brought to bear on the inherently discrete players of biology: genes, cells, proteins, and the other actors in the biological drama. There is simply too much insight to be gained from the continuum approximation not to use it. Until we develop a new form of calculus that works as well for discrete systems as traditional calculus does for continuum ones, the Infinity Principle will continue to guide us in the mathematical modeling of living things.

Determinism and Its Limits

Our next two topics are the rise of nonlinear dynamics and the impact of computers on calculus. I've chosen them because they're so philosophically intriguing in their implications. They could alter the nature of prediction forever and lead to a new era in calculus—and in science more generally—where human insight may begin to fade, although science itself will still go on. To clarify what I mean by this somewhat apocalyptic warning, we need to understand how prediction is possible at all, what it meant classically, and how our classical notions are being revised by discoveries made in the past several decades in studies of nonlinearity, chaos, and complex systems.

Early in the 1800s, the French mathematician and astronomer Pierre Simon Laplace took the determinism of Newton's clockwork universe to its logical extreme. He imagined a godlike intellect (now known as Laplace's demon) that could keep track of all the positions of all the atoms in the universe as well as all the forces acting on them. "If this intellect were also vast enough to submit these data to analysis," he wrote, "nothing would be uncertain and the future just like the past would be present before its eyes."

As the turn of the twentieth century approached, this extreme formulation of the clockwork universe began to seem scientifically and philosophically untenable, for several different reasons. The first came from calculus, and we have Sofia Kovalevskaya to thank for it. Kovalevskaya was born in 1850 and grew up in an aristocratic family in Moscow. When she was eleven she found herself surrounded by calculus, literally—one wall of her bedroom was papered with notes from a calculus course her father had attended in his youth. She later wrote that she "spent whole hours of my childhood in front of that mysterious wall, trying to make out even a single sentence and find the order in which the pages ought to have followed one another." She went on to become the first woman in history to earn a PhD in mathematics.

Although Kovalevskaya showed a flair for mathematics early on, Russian law prevented her from enrolling in college. She entered a marriage of convenience, which caused her much heartache in the years to come but that at least allowed her to travel to Germany, where she impressed several professors as an extraordinary talent. Yet even there, she was not officially allowed to attend their classes. She arranged to study privately with the analyst Karl Weierstrass and, at his recommendation, was awarded a doctorate for solving several outstanding problems in analysis, dynamics, and partial differential equations. She eventually became a full professor at the University of Stockholm and taught there for eight years before dying from influenza at the age of forty-one. In 2009, the Nobel Prize–winning author Alice Munro published a short story about her called "Too Much Happiness."

Kovalevskaya's insights on the limits of determinism came from her work on the dynamics of rigid bodies. A rigid body is a mathematical abstraction of an object that can't be bent or deformed; all of its points are rigidly attached to one another. An example is a spinning top. It's completely solid and composed of infinitely many points and is therefore a more complicated mechanical object than the single point-like particles that Newton had considered. The motion of rigid bodies is important in astronomy and space science for describing phenomena ranging from the chaotic tumbling of Hyperion, a little potato-shaped moon of Saturn, to the regular rotation of a space capsule or satellite.

While studying rigid-body dynamics, Kovalevskaya produced two major results. The first was an example of a spinning top whose motion could be completely analyzed and solved, in the same sense that Newton had solved the two-body problem. Two other such "integrable tops" were already known, but hers was more subtle and surprising.

More important, she proved that no other solvable tops could exist. She had found the last one. All others from then on would be non-integrable, meaning that their dynamics would be impossible to solve with Newtonian-style formulas. It wasn't a matter of insufficient cleverness; she proved that there simply couldn't be any formulas of a certain type (in the jargon, a meromorphic function of time) that could describe the motion of the top forever. In this way, she put limits on what calculus could do. If even a spinning top could defy Laplace's demon, there was no hope—even in principle—of finding a formula for the fate of the universe.

Nonlinearity

The unsolvability that Sofia Kovalevskaya discovered is related to a structural aspect of the equations for a top: the equations are *nonlinear*. The technical meaning of *nonlinear* need not concern us here. For our purposes, all we need is a feel for the distinction between linear and nonlinear systems, which we can get by considering some homey examples from everyday life.

To illustrate what linear systems are like, suppose two people try to weigh themselves by stepping on a scale at the same time, just for the fun of it. Their combined weight will be the sum of their individual weights. That's because a scale is a linear device. The people's weights don't interact with each other or do anything tricky that we need to be aware of. For example, their bodies don't somehow conspire with each other to seem lighter or sabotage each other to seem heavier. They simply add up. On a linear system like a scale, the whole is equal to the sum of the parts. That's the first key property of linearity. The second is that causes are proportional to effects. Imagine pulling on the string of an archer's bow. If it takes a certain amount of force to pull the string back a certain distance, it takes twice as much force to pull it back by twice that distance. Cause and effect are proportional. These two properties—the proportionality between cause and effect, and the equality of the whole to the sum of the parts—are the essence of what it means to be linear.

Yet many things in nature are more complicated than this. Whenever parts of a system interfere or cooperate or compete with each other, there are nonlinear interactions taking place. Most of everyday life is spectacularly nonlinear; if you listen to your two favorite songs at the same time, you won't get double the pleasure. The same goes for consuming alcohol and drugs, where the interaction effects can be deadly. By contrast, peanut butter and jelly are better together. They don't just add up—they synergize.

Nonlinearity is responsible for the richness in the world, for its beauty and complexity and, often, its inscrutability. For example, all of biology is nonlinear; so is sociology. That's why the soft sciences are hard—and the last to be mathematized. Because of nonlinearity, there's nothing soft about them.

The same distinction between linear and nonlinear applies to differential equations, though in a less intuitive fashion. The only thing we need to say is that when differential equations are nonlinear, as they were for Kovalevskaya's tops, they are extremely difficult to analyze. Ever since Newton, mathematicians have avoided nonlinear differential equations wherever possible. They're seen as nasty and recalcitrant.

In contrast, linear differential equations are sweet and docile. Mathematicians love them because they're easy. There's an enormous body of theory for solving them. Indeed, until about the 1980s, the traditional education of an applied mathematician was almost entirely devoted to learning methods to exploit linearity. Years were spent mastering Fourier series and other techniques tailored to linear equations.

The great advantage of linearity is that it allows for reductionist thinking. To solve a linear problem, we can break it down to its simplest parts, solve each part separately, and put the parts back together to get the answer. Fourier solved his heat equation—which was linear—with this reductionist strategy. He broke a complicated temperature distribution into sine waves, figured out how each sine wave would change on its own, then recombined those sine waves to predict how the overall temperature would change along the length of a heated metal rod. The strategy worked because the heat equation is linear. It can be chopped into bits without losing its essence.

Sofia Kovalevskaya helped us understand how different the world appears when we finally face up to nonlinearity. She realized that nonlinearity places limits on human hubris. When a system is nonlinear, its behavior can be impossible to forecast with formulas, even though that behavior is completely determined. In other words, determinism does not imply predictability. It took the motion of a top—a child's plaything—to make us more humble about what we can ever hope to know.

Chaos

In retrospect, we can see more clearly why Newton's head ached when he tried to solve the three-body problem. That problem is inescapably nonlinear, unlike the two-body problem, which can be massaged to become linear. The nonlinearity wasn't caused by the leap from to two to three bodies. It was caused by the structure of the equations themselves. For two gravitating bodies, but not for three or more, the nonlinearity could be eliminated by a felicitous choice of new variables in the differential equations.

It took a long time for the humbling implications of nonlinearity to be fully appreciated. Mathematicians thrashed around for centuries trying to solve the three-body problem, and although progress was made, no one managed to crack it completely. In the late 1800s, the French mathematician Henri Poincaré thought he'd solved it, but he'd made a mistake. When he rectified his error, he still couldn't solve the three-body problem, but he discovered something far more important: the phenomenon that we now call *chaos*.

Chaotic systems are finicky. A little change in how they're started can make a big difference in where they end up. That's because small changes in their initial conditions get magnified exponentially fast. Any tiny error or disturbance snowballs so rapidly that in the long term, the system becomes unpredictable. Chaotic systems are not random—they're deterministic and hence predictable in the short run—but in the long run, they're so sensitive to tiny disturbances that they look effectively random in many respects.

Chaotic systems can be predicted perfectly well up to a time known as the predictability horizon. Before that, the determinism of the system makes it predictable. For example, the horizon of predictability for the entire solar system has been calculated to be about four million years. For times much shorter than that, like the single year it takes our Earth to go around the sun, everything behaves like clockwork. But once we move past a few million years, all bets are off. The subtle gravitational perturbations among all the bodies in the solar system accumulate until we can no longer forecast the system accurately.

The existence of the predictability horizon emerged from Poincaré's work. Before him, it was thought that errors would grow only linearly in time, not exponentially; if you doubled the time, there'd be double the error. With a linear growth of errors, improving the measurements could always keep pace with the desire for longer prediction. But when errors grow *exponentially* fast, a system is said to have sensitive dependence on its initial conditions. Then long-term prediction becomes impossible. This is the philosophically disturbing message of chaos.

It's important to understand what's new about this. People always knew that big complex systems like the weather were hard to predict. The surprise was that something as simple as a spinning top or three gravitating bodies was similarly unpredictable. That was a shocker and another blow to Laplace's naive conflation of determinism with predictability.

On the positive side, vestiges of order exist within chaotic systems because of their deterministic character. Poincaré developed new methods for analyzing nonlinear systems, including chaotic ones, and found ways to extract some of the order hidden within them. Instead of formulas and algebra, he used pictures and geometry. His qualitative approach helped sow the seeds for the modern mathematical fields of topology and dynamical systems. We now have a much better understanding of order and chaos because of his seminal work.

Poincaré's Visual Approach

To give an example of how Poincaré's approach works, consider the oscillations of a simple pendulum of the sort that Galileo studied. Using Newton's law of motion and taking note of the forces that a pendulum experiences as it swings, we can draw an abstract picture showing how the pendulum changes its angle and velocity from moment to moment. That picture is essentially a visual translation of what Newton's law says. There is no new content in the picture beyond what's already in the differential equation. It's just another way of looking at the same information.

The picture looks like a map of a weather pattern traveling across the countryside. On such maps, we see arrows showing the local direction of propagation, which way the weather front will move instant by instant. This is the same kind of information that a differential equation provides. It's also the same kind of information given in dance instructions: put your left foot here, put your right foot there. Such a map is called a graph of a *vector field*. The little arrows on it are vectors showing that if the angle and velocity of the pendulum are currently here, this is where they should go a moment later. The vector-field picture for the pendulum looks like this:

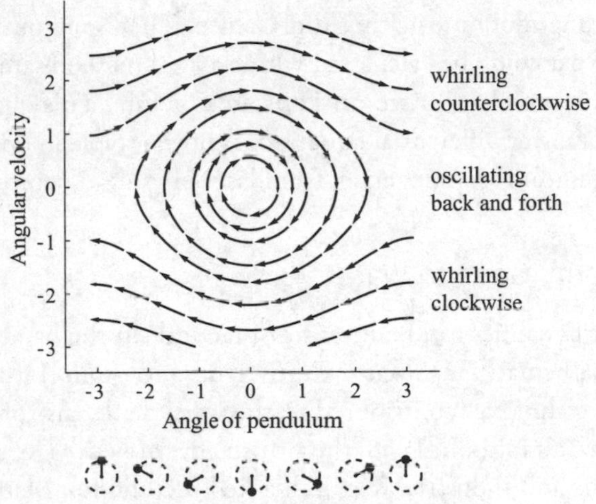

Before we interpret the picture, please understand that it is abstract in the sense that it's not showing a realistic portrait of a pendulum. The pattern of swirling arrows does not resemble a weight hanging from a string. It's not what a photograph of a pendulum would look like. (Cartoons of such snapshots are shown below the vector-field picture to give you a feeling for what it means.) Instead of a realistic depiction of the pendulum, the vector-field picture shows an abstract map of how the state of the pendulum changes from one moment to the next. Each point on the map represents a possible combination of the pendulum's angle and velocity at an instant. The horizontal axis represents the pendulum's angle. The vertical axis represents its velocity. At any moment, a knowledge of those two numbers, angle and velocity, define the dynamical *state* of the pendulum. They provide the information we need to predict what the angle and velocity of the pendulum will be a moment later, and then a moment after that, and so on. All we need to do is follow the arrows.

The swirling arrangement of the arrows near the center corresponds to a simple back-and-forth motion of the pendulum when it is hanging nearly straight down. The wavy structure of the arrows on the top and bottom correspond to a pendulum rotating vigorously over the top, whirling like a propeller. Newton never considered such whirling motions; neither did Galileo. They were outside the realm of what could be calculated with classical methods. Yet whirling motions are plain to see on Poincaré's picture. This qualitative way of looking at differential equations is now a staple in every field where nonlinear dynamics arise, from laser physics to neuroscience.

Nonlinearity Goes to War

Nonlinear dynamics can be intensely practical. In the hands of the British mathematicians Mary Cartwright and John Littlewood, Poincaré's techniques contributed to the wartime defense of Britain against Nazi air raids. In 1938, the British government's Department of Scientific and Industrial Research asked the London Mathematical Society for help with a problem related to top-secret developments in radio detection and ranging, the technology known today as radar. British government engineers working on the project had been perplexed by noisy, erratic oscillations they were observing in their amplifiers, especially when the devices were driven by high-power, high-frequency radio waves. They feared that something might be wrong with their equipment.

The government's call for help caught Cartwright's attention. She had already been studying models of oscillating systems governed by similar "very objectionable-looking differential equations," as she later described them. She and Littlewood went on to discover the source of the erratic oscillations in the radar electronics. The amplifiers were nonlinear, and they could respond chaotically if they were driven too fast and too hard.

Decades later, the physicist Freeman Dyson recalled hearing Cartwright lecture on her work in 1942. He wrote:

> The whole development of radar in World War II depended on high power amplifiers, and it was a matter of life and death to have amplifiers that did what they were supposed to do. The soldiers were plagued with amplifiers that misbehaved, and blamed the manufacturers for their erratic behaviour. Cartwright and Littlewood discovered that the manufacturers were not to blame. The equation itself was to blame.

The insights of Cartwright and Littlewood enabled the government's engineers to work around the problem by operating the amplifiers in regimes where they behaved more predictably. Cartwright was characteristically modest about her contribution. When she read what Dyson had written about her work, she scolded him for making too much of it.

Dame Mary Cartwright passed away in 1998 at the age of ninety-seven. She was the first female mathematician elected to the Royal Society. She left strict instructions that no eulogies were to be given at her memorial service.

The Alliance Between Calculus and Computers

The need to solve differential equations in wartime spurred the development of computers. Mechanical and electronic brains, as they were sometimes called in those days, could be used to calculate the trajectories of rockets and cannon shells under realistic conditions by accounting for complications like air resistance and wind direction. Such information was needed by artillery officers in the field to help them hit their targets. All the necessary ballistic data were computed ahead of time and compiled in standard tables and charts. High-speed computers were essential for this task. In a mathematical simulation, the computers could inch an idealized cannon shell forward on its flight path, one small step at a time, using the appropriate differential equation to update the shell's position and velocity by one small increment after another, proceeding to the solution by brute force through an enormous number of additions. Only a machine could chug forward relentlessly and perform all the necessary additions and multiplications quickly, correctly, and tirelessly.

The legacy of calculus in this endeavor is evident in the names of some of the earliest computers. One was a mechanical device called the Differential Analyzer. Its job was to solve the differential equations needed to compute artillery-firing tables. Another was called ENIAC, for "Electronic Numerical Integrator and Computer." Here the word *integrator* was used in the calculus sense, as in doing integrals or integrating a differential equation. Completed in 1945, ENIAC was one of the first reprogrammable, general-purpose computers. Along with computing firing tables, it also assessed the technical feasibility of a hydrogen bomb.

Although military applications of calculus and nonlinear dynamics stimulated the development of computers, many peacetime uses were found for both the math and the machines. In the 1950s scientists began to use them to solve problems arising in their own disciplines, outside of physics. For example, the British biologists Alan Hodgkin and Andrew Huxley needed computers to help them understand how nerve cells talked to one another and, more specifically, how electrical signals traveled along nerve fibers. They performed painstaking experiments to calculate the flow of sodium and potassium ions across the membrane of a very big and experimentally convenient kind of nerve fiber—the giant axon of a squid—and worked out empirically how those flows depended on the voltage across the membrane and how the voltage was altered by the flowing ions. But what they were not able to do without a computer was calculate the speed and shape of a neural impulse as it traveled down an axon. Calculating its motion required solving a nonlinear partial differential equation for the voltage as a function of time and space. Andrew Huxley solved it over the course of three weeks on a hand-cranked mechanical calculator.

In 1963, Hodgkin and Huxley shared a Nobel Prize for their discoveries about the ionic basis of how nerve cells work. Their approach has been a big inspiration to all those interested in applying mathematics to biology. This is sure to be a growth area for the applications of calculus. Mathematical biology is a no-holds-barred exercise in nonlinear differential equations. With the help of Newton-style analytical methods, Poincaré-style geometric methods, and an unabashed reliance on computers, mathematical biologists are looking for and starting to make headway on the differential equations that govern heart rhythms, the spread of epidemics, the functioning of the immune system, the orchestration of genes, the development of cancer, and many other mysteries of life. We couldn't do any of it without calculus.

Complex Systems and the Curse of High Dimensions

The most serious limitation of Poincaré's approach has to do with the human brain, which can't imagine spaces having more than three dimensions. Natural selection has tuned our nervous systems to perceive up and down, front and back, and left and right, the three directions of ordinary space. Try as we might, we can't picture a fourth dimension, not in the sense of *seeing* it in the mind's eye. With abstract symbols, however, we can try to deal with any number of dimensions. Fermat and Descartes showed us how. Their *xy* plane taught us that numbers could be attached to dimensions. Left and right corresponded to the number *x*. Up and down corresponded to the number *y*. By including more numbers, we could include more dimensions. For three dimensions, *x*, *y*, and *z* sufficed. Why not have four dimensions, or five? There were still plenty of letters left.

You may have heard that time is the fourth dimension. Indeed, in Einstein's special and general theories of relativity, space and time are fused into a single entity, space-time, and represented in a four-dimensional mathematical arena. Roughly speaking, ordinary space gets plotted on the first three axes and time gets plotted on the fourth. This construction can be viewed as a generalization of the two-dimensional *xy* plane of Fermat and Descartes.

But we are not talking about space-time here. The limitation inherent in Poincaré's approach involves a much more abstract arena. It's a generalization of the abstract *state space* we met when we looked at the vector field for a pendulum. In that example, we constructed an abstract space with one axis for the pendulum's angle and another for its velocity. At each instant, the angle and velocity of the swinging pendulum had certain values; hence, at that instant, they corresponded to a single point in the angle-velocity plane. The arrows on that plane (the ones that looked like dance instructions) dictated how the state changed from instant to instant, as determined by Newton's differential equation for the pendulum. By following the arrows, we could forecast how the pendulum would move. Depending on where it started, it could oscillate back and forth or it could whirl over the top. All of that was contained in the picture.

The key thing to realize is that the pendulum's state space had *two* dimensions because *two* variables—the pendulum's angle and its velocity—were necessary and sufficient to predict its future. They gave us exactly the information we needed to predict its angle and velocity an instant later, and an instant after that, on and on into the future. In that sense, the pendulum is an inherently two-dimensional system. It has a two-dimensional state space.

The curse of high dimensions arises when we consider systems more complicated than a pendulum. For example, let's take the problem that gave Newton a headache, the problem of three mutually gravitating bodies. Its state space has eighteen dimensions. To see why, concentrate on one of the bodies. At any instant, it is located somewhere in ordinary three-dimensional physical space. Its location can therefore be specified by three numbers: x, y, z. It can also move in each of those three directions, corresponding to three velocities. So a single body requires *six* pieces of information: three coordinates for its location plus three for its velocity in the different directions. Those six numbers specify where it is and how it's moving. Multiply that six by each of the three bodies in the problem and now you have $6 \times 3 = 18$ dimensions in state space. Thus, in Poincaré's approach, the changing state of a system of three mutually gravitating bodies is represented by a single abstract point moving around in an eighteen-dimensional space. As time passes, the abstract point traces out a trajectory, analogous to the trajectory of a real comet or a cannonball, except this abstract trajectory lives in Poincaré's fantastic arena, the eighteen-dimensional state space of the three-body problem.

When we apply nonlinear dynamics to biology, we often find it necessary to imagine even higher-dimensional spaces. For example, in neuroscience we need to keep track of all the changing concentrations of sodium, potassium, calcium, chloride, and other ions involved in the nerve-membrane equations of Hodgkin and Huxley. Modern versions of their equations can involve hundreds of variables. Those variables represent the changing concentrations of ions in the nerve cell, the changing voltage across the cell membrane, and the membrane's changing ability to conduct the various ions and allow them to pass into the cell or out of it. The abstract state space in this case has hundreds of dimensions, one for each variable—one for potassium concentration, another for sodium concentration, a third for voltage, a fourth for sodium conductance, a fifth for potassium conductance, and so on. At any given instant, all those variables take certain values. The Hodgkin-Huxley equations (or their generalizations) give the variables their dance instructions and tell them how to move on their trajectories. In this way, the dynamics of nerve cells, brain cells, and heart cells can be predicted, sometimes with surprising accuracy, with the help of computers to step the trajectories forward though state space. The fruits of this approach are being used to study neural pathologies and cardiac arrhythmias and to design better defibrillators.

Today, mathematicians regularly think about abstract spaces having arbitrary numbers of dimensions. We speak about *n*-dimensional space, and we have developed geometry and calculus in any number of dimensions. As we saw in chapter 10, Allan Cormack, the inventor of the theory behind CT scanning, wondered how CT would work in four dimensions, purely out of intellectual curiosity. Great things have come from this spirit of pure adventure. When Einstein needed four-dimensional geometry for curved space and time in general relativity, he was pleased to learn it already existed, thanks to Bernhard Riemann, who had created it decades earlier for the purest of mathematical reasons.

So there is a lot to be said for following one's curiosity in mathematics. It often has scientific and practical payoffs that can't be foreseen. It also gives mathematicians great pleasure for its own sake and reveals hidden connections between different parts of mathematics. For all these reasons, the pursuit of higher-dimensional spaces has been a vigorous part of mathematics for the past two hundred years.

However, although we have an abstract system for doing math in high-dimensional spaces, mathematicians still have trouble visualizing them. Actually, let me be more frank—we *can't* visualize them. Our brains just aren't up to it. We aren't wired that way.

That cognitive limitation deals a serious blow to Poincaré's program, at least in dimensions higher than three. His approach to nonlinear dynamics depends on visual intuition. If we can't picture what's going to happen in four or eighteen or a hundred dimensions, his approach can't help us all that much. This has become a big obstacle to progress in the field of *complex systems,* where high-dimensional spaces are exactly what we need to understand if we want to make sense of the thousands of biochemical reactions taking place in a healthy living cell or explain how they go awry in cancer. If we are to have any hope of making sense of cell biology using differential equations, we need to be able to solve those equations with formulas (which Sofia Kovalevskaya showed we cannot) or picture them (which our limited brains won't allow).

So the mathematics of complex nonlinear systems is discouraging. It seems like it will always be hard, if not impossible, for anyone to make headway on the most difficult problems of our time, from the behavior of economies, societies, and cells to the workings of the immune system, genes, brains, and consciousness.

A further difficulty is that we don't even know if some of those systems harbor patterns akin to those uncovered by Kepler and Galileo. Nerve cells apparently do, but what about economies or societies? In many fields, human understanding is still in the pre-Galilean, pre-Keplerian phase. We haven't found the patterns. So how can we find deeper theories that would give insight into those patterns?

Biology and psychology and economics are not Newtonian yet, because they aren't even Galilean and Keplerian. We have a long way to go.

Computers, Artificial Intelligence, and the Mystery of Insight

At this point, the computer triumphalists demand to be heard. With computers, they say, with artificial intelligence, all of these problems will fall. And that may well be true. Computers have long helped us in the study of differential equations, nonlinear dynamics, and complex systems. When Hodgkin and Huxley opened the door in the 1950s to understanding how nerve cells work, they solved their partial differential equations on a hand-cranked machine. When engineers at Boeing designed the 787 Dreamliner in 2011, they used supercomputers to calculate the lift and drag on the plane and figure out how to prevent unwanted vibrations of its wings.

Computers began as calculating machines—literally, *compute-ers*—but they are now much more than that. They have achieved artificial intelligence of a sort. For example, Google Translate now does a surprisingly good job of providing idiomatic translations. And there are medical AI systems that diagnose diseases more accurately than the best human experts.

Still, I don't believe anyone would say that Google Translate has insight into languages or that medical AI systems understand diseases. Could computers ever be insightful? If so, could they share their insights with us about things we really care about, like complex systems, which are central to most of the greatest unsolved problems of science?

To explore the case for and against the possibility of computer insight, consider how computer chess has evolved. In 1997, IBM's chess-playing program Deep Blue managed to beat the reigning human world chess champion, Garry Kasparov, in a six-game match. Although unexpected at the time, there was no great mystery in this achievement. The machine could evaluate two hundred million positions per second. It didn't have insight, but it had raw speed, it never got tired, it never blundered in a calculation, and it never forgot what it was thinking a minute ago. Still, it played like a computer, mechanically and materialistically. It could outcompute Kasparov but it couldn't outthink him. The current generation of the world's strongest chess programs, with intimidating names like Stockfish and Komodo, still play in this inhuman style. They like to capture material. They defend like iron. But although they are far stronger than any human player, they are not creative or insightful.

All that changed with the rise of machine learning. On December 5, 2017, the DeepMind team at Google stunned the chess world with its announcement of a deep-learning program called AlphaZero. The program taught itself chess by playing millions of games against itself and learning from its mistakes. In a matter of hours, it became the best chess player in history. Not only could it easily defeat all the best human masters (it didn't even bother to try), it crushed the reigning computer world champion of chess. In a hundred-game match against Stockfish, a truly formidable program, AlphaZero scored twenty-eight wins and seventy-two draws. It didn't lose a single game.

The scariest point is that AlphaZero showed insight. It played like no computer ever has, intuitively and beautifully, with a romantic, attacking style. It played gambits and took risks. In some games it paralyzed Stockfish and toyed with it. It seemed malevolent and sadistic. And it was creative beyond words, playing moves no grandmaster or computer would ever dream of making. It had the spirit of a human and the power of a machine. It was humankind's first glimpse of a terrifying new kind of intelligence.

Suppose we could unleash AlphaZero or something like it—let's call it AlphaInfinity—on the greatest unsolved problems in theoretical science, problems of immunology and cancer biology and consciousness. To continue the fantasy, suppose that Galilean and Keplerian patterns exist in these phenomena and are ripe for the picking, but only by an intelligence far superior to ours. Assuming that such laws exist, would this superhuman intelligence be able to work them out? I don't know. No one knows. And it all may be moot, because such laws may not even exist.

But if they do, and if AlphaInfinity could find them, it would seem like an oracle to us. We'd sit at its feet and listen to it. We wouldn't understand why it was always right or even what it was saying, but we could check its calculations against experiments or observations, and it would seem to know everything. We would be reduced to spectators, gaping in wonder and confusion. Even if it could explain itself, we wouldn't be able to follow its reasoning. At that moment, the age of insight that began with Newton would come to a close, at least for humanity, and a new age of insight would begin.

Science fiction? Perhaps. But I think a scenario like this is not out of the question. In parts of mathematics and science, we are already experiencing the dusk of insight. There are theorems that have been proved by computers, yet no human being can understand the proof. The theorems are correct but we have no insight into why. And at this point, the machines cannot explain themselves.

Consider the famous long-standing math problem called the four-color map theorem. It says that under certain reasonable constraints, any map of contiguous countries can always be colored with just four colors such that no two neighboring countries are colored the same. (Look at a typical map of Europe or Africa or any other continent besides Australia and you'll see what I mean.) The four-color theorem was proved in 1977 with the help of a computer, but no human being could check all the steps in the argument. Although the proof has been validated and simplified since then, there are parts of it that unavoidably entail brute-force computation, like the way computers used to play chess before AlphaZero. When this proof came out, many working mathematicians were cranky about it. They already believed the four-color theorem. They did not need any assurance that it was true. They wanted to understand *why* it was true, and this proof didn't help.

Likewise, consider a four-hundred-year-old geometry problem posed by Johannes Kepler. It asks for the densest way to pack equal-size spheres in three dimensions, akin to the problem faced by grocers when they pack oranges in a crate. Would it be most efficient to stack the spheres in identical layers, one directly on top of another?

Or would it be better to stagger the layers so that each sphere nestles in the hollow formed by four others beneath it, the same way grocers stack oranges? If so, is that packing the best possible one, or could some other, possibly irregular, packing arrangement be denser? Kepler's conjecture was that the grocers' packing is the best. But this wasn't proved until 1998. Thomas Hales, with the help of his student Samuel Ferguson and 180,000 lines of computer code, reduced the calculation to a large but finite number of cases. Then, with the help of brute-force computation and ingenious algorithms, his program verified the conjecture. The mathematical community shrugged. We now know that the Kepler conjecture is true but we still don't understand why. We don't have insight. Nor could Hales's computer explain it to us.

But what about when we unleash AlphaInfinity on such problems? A machine like that would come up with beautiful proofs, as beautiful as the chess games that AlphaZero played against Stockfish. Its proofs would be intuitive and elegant. They would be, in the words of the Hungarian mathematician Paul Erdős, proofs straight from the Book. Erdős imagined that God kept a book with all the best proofs in it. Saying that a proof was straight from the Book was the highest possible praise. It meant that the proof revealed *why* a theorem was true and didn't merely bludgeon the reader into accepting it with some ugly, difficult argument. I can imagine a day, not too far in the future, when artificial intelligence will give us proofs from the Book. What will calculus be like then, and what will medicine be like, and sociology, and politics?

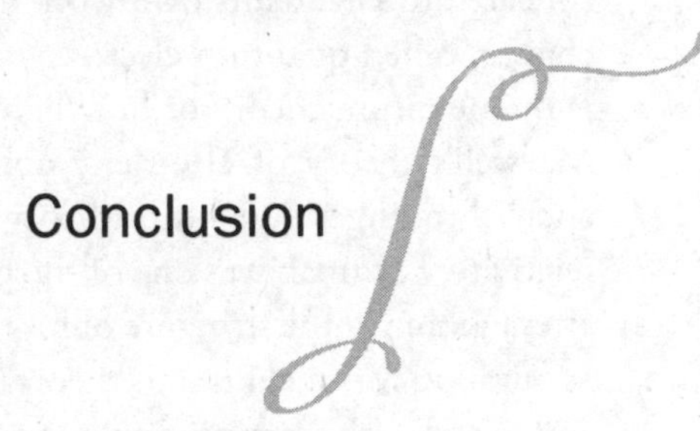

Conclusion

By wielding infinity in just the right way, calculus can unlock the secrets of the universe. We've seen that happen again and again, but it still seems almost miraculous. A system of reasoning humans invented is somehow in tune with the harmony of nature. It's reliable not just at the scales where it was invented—at the everyday scales of ordinary life, with its spinning tops and its bowls of soup—but also at the smallest scales of atoms and at the grandest scales of the cosmos. So it can't just be a trick of circular reasoning. It's not that we're stuffing things into calculus that we already know, and calculus is handing them back to us; calculus tells us about things we've never seen, never could see, and never will see. In some cases, it tells us about things that never existed but could—if only we had the wit to conjure them.

This, to me, is the greatest mystery of all: Why is the universe comprehensible, and why is calculus in sync with it? I have no answer, but I hope you'll agree it's worth contemplating. In that spirit, let me take you to the Twilight Zone for three final examples of the eerie effectiveness of calculus.

Eight Decimal Places

The first example takes us back to where we started, with Richard Feynman's quip that calculus is the language God talks. The example is related to Feynman's own work on an extension of quantum mechanics called quantum electrodynamics, or QED for short. QED is the quantum theory of how light and matter interact. It merges Maxwell's theory of electricity and magnetism with Heisenberg's and Schrödinger's quantum theory and Einstein's special theory of relativity. Feynman was one of the principal architects of QED, and after looking at the structure of his theory, I can see why he had such admiration for calculus. His theory is chock-full of it, both in tactics and in style. It's teeming with power series, integrals, and differential equations and includes plenty of hijinks with infinity.

More important, it's the most accurate theory anyone has ever devised . . . about anything. With the help of computers, physicists are still busy summing the series that arise in QED, using what are known as Feynman diagrams, to make predictions about the properties of electrons and other particles. By comparing those predictions to extremely precise experimental measurements, they've shown that the theory agrees with reality to eight decimal places, better than *one part in a hundred million*.

This is a fancy way of saying that the theory is essentially right. It's always hard to find helpful analogies to make sense of such big numbers, but let me try putting it like this: a hundred million seconds equals 3.17 years, so getting something right to within one part in a hundred million is like planning to snap your fingers exactly 3.17 years from now and timing it right to the nearest second —without the help of a clock or an alarm.

There's something astonishing about this, philosophically speaking. The differential equations and integrals of quantum electrodynamics are creations of the human mind. They are based on experiments and observations, certainly, so they have reality built into them to that extent. Yet they are products of the imagination nonetheless. They are not slavish imitations of reality. They are inventions. And what is so astonishing is that by making certain scribbles on paper and doing certain calculations with methods analogous to those developed by Newton and Leibniz but souped up for the twenty-first century, we can predict nature's innermost properties and get them right to eight decimal places. Nothing that humanity has ever predicted is as accurate as the predictions of quantum electrodynamics.

I think this is worth mentioning because it puts the lie to the line you sometimes hear, that science is like faith and other belief systems, that it has no special claim on truth. Come on. Any theory that agrees to one part in a hundred million is not just a matter of faith or somebody's opinion. It didn't have to match to eight decimal places. Plenty of theories in physics have turned out to be wrong. Not this one. Not yet, at least. No doubt it's a little bit off, as every theory always is, but it sure comes close to the truth.

Summoning the Positron

The second example of the eerie effectiveness of calculus has to do with an earlier extension of quantum mechanics. In 1928, the British physicist Paul Dirac tried to find a way to reconcile Einstein's special theory of relativity with the governing principles of quantum mechanics as applied to an electron moving near the speed of light. He came up with a theory that struck him as beautiful. He chose it largely on aesthetic grounds. He had no particular empirical evidence for the theory, just an artistic sense that its beauty was a sign of its correctness. Those constraints alone—compatibility with relativity and quantum mechanics along with mathematical elegance—tied his hands to a large extent. After struggling with various theories, he found one that matched all his aesthetic desiderata. The theory, in other words, was guided by a quest for harmony. And like any good scientist, Dirac sought to test his theory by extracting predictions from it. For him, as a theoretical physicist, that meant using calculus.

When he solved his differential equation, now known as the Dirac equation, and kept analyzing it over the next few years, he found that it made several startling predictions. One was that *antimatter should exist.* There should be, in other words, a particle equivalent to an electron but with a positive charge. At first he thought that particle might be a proton, but a proton had too much mass; the particle he predicted was about two thousand times smaller than a proton.

No such positively charged particle that wispy had ever been seen. Yet his equation was predicting it. Dirac called it an anti-electron. In 1931 he published a paper in which he predicted that when this still-unobserved particle collided with an electron, the two would annihilate each other. "This new development requires no change whatever in the formalism when expressed in terms of abstract symbols," he wrote, and he added dryly, "Under these circumstances one would be surprised if Nature had made no use of it."

The next year, an experimental physicist named Carl Anderson saw an odd track in his cloud chamber when he was studying cosmic rays. Some sort of particle was coiling like an electron but curving in the opposite direction, as if it had a positive charge. He was unaware of Dirac's prediction, but he got the gist of what he was seeing. When Anderson published a paper about it in 1932, his editor suggested he call it a positron. The name stuck. Dirac won a Nobel Prize for his equation the next year; Anderson won for the positron in 1936.

In the years since then, positrons have been put to work saving lives. They underlie PET scans (*PET* stands for *positron emission tomography*), a form of medical imaging that allows doctors to see regions of abnormal metabolic activity in soft tissues in the brain or other organs. In a noninvasive fashion that requires no surgery or other dangerous intrusions into the skull, PET scans can help locate brain tumors and detect the amyloid plaques associated with Alzheimer's disease.

So here is another sterling example of calculus as the handmaiden to something marvelously practical and important. Because calculus is the language of the universe as well as the logical engine for extracting its secrets, Dirac was able to write down a differential equation for the electron that told him something new and true and beautiful about nature. It led him to conjure up a new particle and realize that it ought to exist. Logic and beauty demanded it. But not on their own—they had to align with known facts and mesh with known theories. When all of that was stirred into the pot, it was almost as if the symbols themselves brought the positron into existence.

The Mystery of a Comprehensible Universe

For our third example of the eerie effectiveness of calculus, it seems appropriate to end our journey in the company of Albert Einstein. He embodied so many of the themes we've touched on: a reverence for the harmony of nature, a conviction that mathematics is a triumph of the imagination, a sense of wonder at the comprehensibility of the universe.

Nowhere are these themes more clearly visible than in his general theory of relativity. In this theory, his magnum opus, Einstein overturned Newton's conceptions of space and time and redefined the relationship between matter and gravity. To Einstein, gravity was no longer a force acting instantaneously at a distance. Instead, it was an almost palpable thing, a warp in the fabric of the universe, a manifestation of the curvature of space and time. Curvature—an idea that goes back to the birth of calculus, to the ancient fascination with curved lines and curved surfaces—in Einstein's hands became a property not just of shapes but of space itself. It's as if the *xy* plane of Fermat and Descartes took on a life of its own. Instead of being an arena for the drama, space became an actor in its own right. In Einstein's theory, matter tells space-time how to curve, while curvature tells matter how to move. The dance between them makes the theory nonlinear.

And we know what that means: Understanding what the equations imply is bound to be difficult. To this day, the nonlinear equations of general relativity conceal many secrets. Einstein was able to excavate some of them through his mathematical skill and doggedness. He predicted, for example, that starlight would bend as it passed around the sun on its way to our planet, a prediction that was confirmed during a solar eclipse in 1919 and that made Einstein an international sensation, front-page news in the *New York Times*.

The theory also predicted that gravity could have a strange effect on time: The passage of time could speed up or slow down as an object moves through a gravitational field. Bizarre as this sounds, it really does occur. It needs to be taken into account in the satellites of the global positioning system as they move high above the

Earth. The gravitational field is weaker up there, which reduces the curvature of space-time and causes clocks to run faster than they do on the ground. Without correcting for this effect, the clocks aboard the GPS satellites wouldn't keep accurate time. They'd get ahead of ground-based clocks by about 45 microseconds per day. That may not sound like much, but keep in mind that the whole global positioning system requires nanosecond accuracy to work properly, and 45 microseconds is 45,000 nanoseconds. Without the correction for general relativity, errors in global positions would accumulate at about ten kilometers each day, and the whole system would become worthless for navigation in a matter of minutes.

The differential equations of general relativity make several other predictions, such as the expansion of the universe and the existence of black holes. All seemed outlandish when they were predicted, yet all have turned out to be true.

The 2017 Nobel Prize in Physics was awarded for the detection of another outrageous effect predicted by general relativity: gravitational waves. The theory showed that a pair of black holes rotating around each other would swirl the space-time around them, stretching it and squeezing it rhythmically. The resulting disturbance in the fabric of space-time was predicted to spread outward like a ripple moving at the speed of light. Einstein doubted that it would ever be possible to measure such a wave; he worried it might be a mathematical illusion. The achievement of the team that won the Nobel Prize was to design and build the most sensitive detector ever made. On September 14, 2015, their apparatus detected a space-time tremor a thousand times smaller than the diameter of a proton. For comparison, that's like tweaking the distance to the nearest star by the width of a human hair.

It's a clear winter night as I write these last words. I've stepped out to look at the sky. With the stars up above and the blackness of space, I can't avoid feeling awe.

How could we, *Homo sapiens,* an insignificant species on an insignificant planet adrift in a middleweight galaxy, have managed to predict how space and time would tremble after two black holes collided in the vastness of the universe a billion light-years away?

We knew what that wave should sound like before it got here. And, courtesy of calculus, computers, and Einstein, we were right.

That gravitational wave was the faintest whisper ever heard. That soft little wave had been headed our way from before we were primates, before we were mammals, from a time in our microbial past. When it arrived that day in 2015, because we were listening —and because we knew calculus—we understood what the soft whisper meant.

Acknowledgments

Writing about calculus for the general public has been a wonderful challenge and a lot of fun. I've been in love with calculus ever since I first learned it in high school and have long dreamed of sharing that love with a wide readership, but I somehow never managed to get around to it. Something would always come up. There were research papers to write, grad students to mentor, classes to prepare, kids to raise, and a dog to walk. Then, about two years ago, it dawned on me that my age (like yours, I bet) was increasing at a rate of one year per year, and so it seemed like that was as good a time as any to try to share the joy of calculus with everyone. My first acknowledgment is therefore to you, dear reader. I've been imagining you for decades. Thanks for being here now.

As it turned out, writing the book I always meant to write has been harder than expected. This shouldn't have been a surprise, but it was. I've been immersed in calculus for so long that it's been hard to see it through the eyes of a newcomer. Fortunately, I've been helped by some very smart, generous, and patient people who didn't have the foggiest idea what calculus was or why it mattered and who certainly didn't spend every waking minute thinking about math the way my colleagues and I do.

Thank you to my literary agent, Katinka Matson. A long time ago, when I offhandedly mentioned that calculus was one of the greatest ideas anyone ever had, you said that was a book you'd like to read. Well, here it is. Thank you so much for believing in me and in this project.

I've been blessed to work with two brilliant editors, Eamon Dolan and Alex Littlefield. Eamon, I can't begin to thank you enough. Working with you was fantastic, from beginning to end. You were the reader I always had in mind: whip-smart, a little bit skeptical, curious, and eager to be thrilled. Best of all, you found the structure in the story before I did and guided me with a firm but gentle hand. I forgive you for asking me to do draft after draft, because you made the book better every time. Truly, I couldn't have done it without you. Alex, thank you for shepherding this manuscript to the finish line and for being such a pleasure to work with in every way.

Speaking of pleasure, what a treat it is to be copyedited by Tracy Roe. Tracy, it almost makes me want to write another book, just for the good-natured education you give me every time we work together.

Editorial assistant Rosemary McGuinness, thank you for your cheerfulness, efficiency, and attention to detail. And thanks to everyone at Houghton Mifflin Harcourt for all your hard work and for being such great team players. I feel lucky to work with you all.

Margy Nelson did the illustrations for this book, just as she has for my others. Thanks as always for your sense of whimsy and collaborative spirit.

I'm grateful to my colleagues Michael Barany, Bill Dunham, Paul Ginsparg, and Manil Suri, who kindly read sections of the book or entire drafts, improved my phrasing, corrected my errors (who knew that there were two Mercators?), and offered helpful suggestions in the jovially nitpicking fashion that every academic hopes for. Michael, I learned so much from your comments and wish I'd shown you the book earlier. Bill, you are a hero. Paul, you are what you always are (and the best at it). Manil, thank you for reading my first draft so carefully and best of luck with your new book, which I can't wait to read in print.

Tom Gilovich, Herbert Hui, and Linda Woodard: Thank you for being such good friends. You let me blabber on about the book for close to two years as it was hatching and never wavered in your encouragement or, as far as I could tell, your attention. Alan Perelson and John Stillwell: I admire your work enormously and feel honored that you shared your thoughts with me about this book. Thank you, too, to Rodrigo Tetsuo Argenton, Tony DeRose, Peter Schröder, Tunç Tezel, and Stefan Zachow, who allowed me to discuss their research and reproduce their published illustrations.

To Murray: You've heard me say it a million times, and even though you don't understand what it means, I know you get the drift. Who's a good boy? You are.

Finally, thank you to my wife, Carole, and daughters, Jo and Leah, for all your love and support and for putting up with my distracted air, which must have been even more annoying than usual. Zeno's paradox about walking halfway to the wall took on new meaning in our household when it seemed like this project was approaching completion but never quite getting there. I am so grateful to you all for your patience and love you very much.

Steven Strogatz
Ithaca, New York

Illustration Credits

page 51 Peter Schröder
page 52 Entertainment Pictures / Alamy
page 53 Entertainment Pictures / Alamy
page 54 Stefan Zachow, Zuse Institute Berlin (ZIB)
page 56 Stefan Zachow, Zuse Institute Berlin (ZIB)
page 61 Tunç Tezel
page 160 WENN Ltd / Alamy
page 182 Reproduced by kind permission of the Syndics of Cambridge University Library. MS-ADD-04000-000-00259.tif (MS Add. 4000, page 124r).
page 240 © Thomas Jefferson Foundation at Monticello
page 260 Image reproduced by kind permission of Rodrigo Tetsuo Argenton

Notes

Introduction

page

vii *"It's the language God talks":* Wouk, *The Language God Talks,* 5.

universe is deeply mathematical: For physics perspectives, see Barrow and Tipler, *Anthropic Cosmological Principle;* Rees, *Just Six Numbers;* Davies, *The Goldilocks Enigma;* Livio, *Is God a Mathematician?;* Tegmark, *Our Mathematical Universe;* and Carroll, *The Big Picture*. For a philosophy perspective, see Simon Friederich, "Fine-Tuning," *Stanford Encyclopedia of Philosophy,* https://plato.stanford.edu/archives/spr2018/entries/fine-tuning/.

viii *answer to the ultimate question of life, the universe, and everything:* Adams, *Hitchhiker's Guide,* and Gill, *Douglas Adams' Amazingly Accurate Answer*.

ix *"a mathematical ignoramus like me":* Wouk, *The Language God Talks,* 6.

x *tell it differently:* For historical treatments, see Boyer, *The History of the Calculus,* and Grattan-Guinness, *From the Calculus.* Dunham, *The Calculus Gallery;* Edwards, *The Historical Development;* and Simmons, *Calculus Gems,* tell the story of calculus by walking us through some of its most beautiful problems and solutions.

To be an applied mathematician: Stewart, *In Pursuit of the Unknown;* Higham et al., *The Princeton Companion;* and Goriely, *Applied Mathematics,* convey the spirit, breadth, and vitality of applied mathematics.

pristine, hermetically sealed world: Kline, *Mathematics in Western Culture,* and Newman, *The World of Mathematics*, connect math to the wider culture. I spent many hours in high school reading these two masterpieces.

xi *electricity and magnetism:* For the mathematics and physics, see Maxwell, "On Physical Lines of Force," and Purcell, *Electricity and Magnetism.* For concepts and history, see Kline, *Mathematics in Western Culture,* 304–21; Schaffer, "The Laird of Physics"; and Stewart, *In Pursuit of the Unknown*, chapter 11. For a biography of Maxwell and Faraday, see Forbes and Mahon, *Faraday, Maxwell.*

wave equation: Stewart, *In Pursuit of the Unknown*, chapter 8.

xiii *"The eternal mystery of the world":* Einstein, *Physics and Reality,* 51. This aphorism is often rephrased as "The most incomprehensible thing about the universe is that it is comprehensible." For further examples of Einstein quotes both real and imaginary, see Calaprice, *The Ultimate Quotable Einstein,* and Robinson, "Einstein Said That."

"Unreasonable Effectiveness of Mathematics": Wigner, "The Unreasonable Effectiveness"; Hamming, "The Unreasonable Effectiveness"; and Livio, *Is God a Mathematician?*

Pythagoras: Asimov, *Asimov's Biographical Encyclopedia*, 4–5; Burkert, *Lore and Science;* Guthrie, *Pythagorean Sourcebook;* and C. Huffman, "Pythagoras," https://plato.stanford.edu/archives/sum2014/entries/pythagoras/. Martínez, in *Cult of Pythagoras* and *Science Secrets*, debunks many of the myths about Pythagoras with a light touch and devastating humor.

xiv *the Pythagoreans:* Katz, *History of Mathematics,* 48–51, and Burton, *History of Mathematics*, section 3.2, discuss Pythagorean mathematics and philosophy.

xxii *stimulated emission:* Ball, "A Century Ago Einstein Sparked," and Pais, *Subtle Is the Lord.* The original paper is Einstein, "Zur Quantentheorie der Strahlung."

1. Infinity

1 *beginnings of mathematics:* Burton, *History of Mathematics*, and Katz, *History of Mathematics,* provide gentle yet authoritative introductions to the history of mathematics from ancient times to the twentieth century. At a more advanced mathematical level, Stillwell, *Mathematics and Its History,* is excellent. For a wide-ranging humanistic treatment with a healthy dose of crotchety opinion thrown in, Kline, *Mathematics in Western Culture*, is delightful.

3 *an outgrowth of geometry:* See section 4.5 of Burton, *History of Mathematics;* chapters 2 and 3 in Katz, *History of Mathematics;* and chapter 4 in Stillwell, *Mathematics and Its History.*

4 *area of a circle:* Katz, *History of Mathematics,* section 1.5, discusses ancient estimates of the area of a circle made by various cultures around the world. The first proof of the formula was given by Archimedes using the

method of exhaustion; see Dunham, *Journey Through Genius*, chapter 4, and Heath, *The Works of Archimedes*, 91–93.

15 *Aristotle:* Henry Mendell, "Aristotle and Mathematics," *Stanford Encyclopedia of Philosophy*, https://plato.stanford.edu/archives/spr2017/entries/aristotle-mathematics/.

completed infinity: Katz, *History of Mathematics*, 56, and Stillwell, *Mathematics and Its History*, 54, discuss Aristotle's distinction between completed (or actual) infinity and potential infinity.

16 *Giordano Bruno:* Drawing on new evidence, Martínez, *Burned Alive*, argues that Bruno was executed for his cosmology, not his theology. Also see A. A. Martínez, "Was Giordano Bruno Burned at the Stake for Believing in Exoplanets?," *Scientific American* (2018), https://blogs.scientificamerican.com/observations/was-giordano-bruno-burned-at-the-stake-for-believing-in-exoplanets/. See also D. Knox, "Giordano Bruno," *Stanford Encyclopedia of Philosophy*, https://plato.stanford.edu/entries/bruno/.

immeasurably subtle and profound: Russell's essay on Zeno and infinity is "Mathematics and the Metaphysicians," reprinted in Newman, *The World of Mathematics*, vol. 3, 1576–90.

17 *Zeno's paradoxes:* Mazur, *Zeno's Paradox*. See also Burton, *History of Mathematics*, 101–2; Katz, *History of Mathematics*, section 2.3.3; Stillwell, *Mathematics and Its History*, 54; John Palmer, "Zeno of Elea," *Stanford Encyclopedia of Philosophy*, https://plato.stanford.edu/archives/spr2017/entries/zeno-elea/; and Nick Huggett, "Zeno's Paradoxes," *Stanford Encyclopedia of Philosophy*, https://plato.stanford.edu/entries/paradox-zeno/.

21 *Quantum mechanics:* Greene, *The Elegant Universe*, chapters 4 and 5.

22 *Schrödinger's equation:* Stewart, *In Pursuit of the Unknown*, chapter 14.

23 *Planck length:* Greene, *The Elegant Universe*, 127–31, explains why physicists believe that space dissolves into quantum foam at the ultramicroscopic scale of the Planck length. For philosophy, see S. Weinstein and D. Rickles, "Quantum Gravity," *Stanford Encyclopedia of Philosophy*, https://plato.stanford.edu/entries/quantum-gravity/.

2. The Man Who Harnessed Infinity

27 *Archimedes:* For his life, see Netz and Noel, *The Archimedes Codex*, and C. Rorres, "Archimedes," https://www.math.nyu.edu/~crorres/Archimedes/contents.html. For a scholarly biography, see M. Clagett, "Archimedes," in Gillispie, *Complete Dictionary*, vol. 1, with amendments by F. Acerbi in vol. 19. For Archimedes's mathematics, Stein, *Archimedes*, and Edwards,

The Historical Development, chapter 2, are both outstanding, but see also Katz, *History of Mathematics,* sections 3.1–3.3, and Burton, *History of Mathematics*, section 4.5. A scholarly collection of Archimedes's work is Heath, *The Works of Archimedes*.

stories about him: Martínez, *Cult of Pythagoras*, chapter 4, traces the evolution of the many legends about Archimedes, including the comical Eureka tale and the tragic story of Archimedes's death at the hands of a Roman soldier during the siege of Syracuse in 212 BCE. While it seems likely that Archimedes was killed during the siege, there's no reason to believe his final words were "Don't disturb my circles!"

Plutarch: The Plutarch quotes are from John Dryden's translation of Plutarch's *Marcellus*, available online at http://classics.mit.edu/Plutarch/marcellu.html. The specific passages about Archimedes and the siege of Syracuse are also available at https://www.math.nyu.edu/~crorres/Archimedes/Siege/Plutarch.html.

"*made him forget his food*"*:* http://classics.mit.edu/Plutarch/marcellu.html.

"carried by absolute violence to bathe": Ibid.

Vitruvius: The Eureka story, as first told by Vitruvius, is available in Latin and English at https://www.math.nyu.edu/~crorres/Archimedes/Crown/Vitruvius.html. That site also includes a children's version of the story by the acclaimed writer James Baldwin, taken from *Thirty More Famous Stories Retold* (New York: American Book Company, 1905). Unfortunately, Baldwin and Vitruvius oversimplify Archimedes's solution to the problem of the king's golden crown. Rorres offers a more plausible account at https://www.math.nyu.edu/~crorres/Archimedes/Crown/CrownIntro.html, along with Galileo's guess regarding how Archimedes might have solved it (https://www.math.nyu.edu/~crorres/Archimedes/Crown/bilancetta.html).

28 *"A ship was frequently lifted up":* http://classics.mit.edu/Plutarch/marcellu.html.

29 *estimate pi:* Stein, *Archimedes,* chapter 11, shows in detail how Archimedes did it. Be prepared for some hairy arithmetic.

33 *existence of irrational numbers:* No one really knows who first proved that the square root of 2 is irrational or, equivalently, that the diagonal of a square is incommensurable with its side. There's an irresistible old yarn that a Pythagorean named Hippasus was drowned at sea for it. Martínez, *Cult of Pythagoras*, chapter 2, tracks down the origin of this myth and debunks it. So does the American filmmaker Errol Morris in a long and wonderfully quirky essay in the *New York Times;* see Errol Morris, "The Ashtray:

Hippasus of Metapontum (Part 3)," *New York Times*, March 8, 2001, https://opinionator.blogs.nytimes.com/2011/03/08/the-ashtray-hippasus-of-metapontum-part-3/.

35 Quadrature of the Parabola: A translation of Archimedes's original text is in Heath, *The Works of Archimedes*, 233–52. For the details I glossed over in the triangular-shard argument, see Edwards, *The Historical Development*, 35–39; Stein, *Archimedes*, chapter 7; Laubenbacher and Pengelley, *Mathematical Expeditions*, section 3.2; and Stillwell, *Mathematics and Its History*, section 4.4. There are also many treatments available on the internet. One of the clearest is by Mark Reeder at https://www2.bc.edu/mark-reeder/1103quadparab.pdf. Another is by R.A.G. Seely at http://www.math.mcgill.ca/rags/JAC/NYB/exhaustion2.pdf. As an alternative, Simmons, *Calculus Gems*, section B.3, uses an analytic geometry approach that you may find easier to follow.

40 *"When you have eliminated the impossible"*: Arthur Conan Doyle, *The Sign of the Four* (London: Spencer Blackett, 1890), https://www.gutenberg.org/files/2097/2097-h/2097-h.htm.

42 *The Method:* For the original text, see Heath, *The Works of Archimedes*, 326 and following. For the application of the Method to the quadrature of the parabola, see Laubenbacher and Pengelley, *Mathematical Expeditions*, section 3.3, and Netz and Noel, *The Archimedes Codex,* 150–57. For the application of the Method to several other problems about areas, volumes, and centers of gravity, see Stein, *Archimedes,* chapter 5, and Edwards, *The Historical Development,* 68–74.

"does not furnish an actual demonstration": Quoted in Stein, *Archimedes*, 33.

"theorems which have not yet fallen to our share": Quoted in Netz and Noel, *The Archimedes Codex,* 66–67.

47 *"made up of all the parallel lines":* Heath, *The Works of Archimedes*, 17.

"drawn inside the curve": Dijksterhuis, *Archimedes*, 317. Dijksterhuis argues, as I have here, that the Method aired some dirty laundry. It revealed that the use of completed infinity "had only been banished from the published treatises," but that didn't stop Archimedes from using it in private. As Dijksterhuis put it, "In the workshop of the producing mathematician," arguments based on completed infinity "held undiminished sway."

"a sort of indication": Heath, *The Works of Archimedes*, 17.

48 *volume of a sphere:* Stein, *Archimedes*, 39–41.

49 *"inherent in the figures":* Heath, *The Works of Archimedes*, 1.

50 *Archimedes Palimpsest:* See Netz and Noel, *The Archimedes Codex;* the authors tell the story of the lost manuscript and its rediscovery with great

panache. There was also a terrific *Nova* episode about it, and the accompanying website offers timelines, interviews, and interactive tools; see http://www.pbs.org/wgbh/nova/archimedes/. See also Stein, *Archimedes*, chapter 4.

Archimedes's legacy: Rorres, *Archimedes in the Twenty-First Century*.

computer-animated movies: For the math behind computer-generated movies and video, see McAdams et al., "Crashing Waves."

triangulations of a mannequin's head: Zorin and Schröder, "Subdivision for Modeling," 18.

51 *Shrek:* DreamWorks, "Why Computer Animation Looks So Darn Real," July 9, 2012, https://mashable.com/2012/07/09/animation-history-tech/#uYHyf6hO.Zq3.

forty-five million polygons: Shrek, production information, http://cinema.com/articles/463/shrek-production-information.phtml.

Avatar: "NVIDIA Collaborates with Weta to Accelerate Visual Effects for Avatar," http://www.nvidia.com/object/wetadigital_avatar.html, and Barbara Robertson, "How Weta Digital Handled Avatar," *Studio Daily*, January 5, 2010, http://www.studiodaily.com/2010/01/how-weta-digital-handled-avatar/.

first movie to use polygons by the billions: "NVIDIA Collaborates with Weta."

Toy Story: Burr Snider, "The Toy Story Story," *Wired*, December 1, 1995, https://www.wired.com/1995/12/toy-story/.

"more PhDs working on this film": Ibid.

Geri's Game: Ian Failes, "'Geri's Game' Turns 20: Director Jan Pinkava Reflects on the Game-Changing Pixar Short," November 25, 2017, https://www.cartoonbrew.com/cgi/geris-game-turns-20-director-jan-pinkava-reflects-game-changing-pixar-short-154646.html. The movie is on YouTube at https://www.youtube.com/watch?v=gLQG3sORAJQ (original soundtrack) and https://www.youtube.com/watch?v=9IYRC7g2ICg (modified soundtrack).

52 *subdivision process:* DeRose et al., "Subdivision Surfaces." Explore subdivision surfaces for computer animation interactively at Khan Academy in collaboration with Pixar at https://www.khanacademy.org/partner-content/pixar/modeling-character. Students and their teachers might also enjoy trying the other lessons offered in "Pixar in a Box," a "behind-the-scenes look at how Pixar artists do their jobs," at https://www.khanacademy.org/partner-content/pixar. It's a great way to see how math is being used to make movies these days.

53 *double chin:* DreamWorks, "Why Computer Animation Looks So Darn Real."

facial surgery: Deuflhard et al., "Mathematics in Facial Surgery"; Zachow et al., "Computer-Assisted Planning"; and Zachow, "Computational Planning."

56 *Archimedean screw:* Rorres, *Archimedes in the Twenty-First Century,* chapter 6, and https://www.math.nyu.edu/~crorres/Archimedes/Screw/Applications.html.

57 *Archimedes was silent:* In fairness, Archimedes did do one study related to motion, though it was an artificial form of motion motivated by mathematics rather than physics. See his essay "On Spirals," reproduced in Heath, *The Works of Archimedes,* 151–88. Here Archimedes anticipated the modern ideas of polar coordinates and parametric equations for a point moving in a plane. Specifically, he considered a point moving uniformly in the radial direction away from the origin at the same time as the radial ray rotated uniformly, and he showed that the trajectory of the moving point is the curve now known as an Archimedean spiral. Then, by summing $1^2 + 2^2 + \cdots + n^2$ and applying the method of exhaustion, he found the area bounded by one loop of the spiral and the radial ray. See Stein, *Archimedes,* chapter 9; Edwards, *The Historical Development,* 54–62; and Katz, *History of Mathematics,* 114–15.

3. Discovering the Laws of Motion

60 *"this grand book":* Galileo, *The Assayer* (1623). Selections translated by Stillman Drake, *Discoveries and Opinions of Galileo* (New York: Doubleday, 1957), 237–38, https://www.princeton.edu/~hos/h291/assayer.htm.

"coeternal with the divine mind": Johannes Kepler, *The Harmony of the World,* translated by E. J. Aiton, A. M. Duncan, and J. V. Field, *Memoirs of the American Philosophical Society* 209 (1997): 304.

"supplied God with patterns": Ibid.

Plato had taught: Plato, *Republic* (Hertfordshire: Wordsworth, 1997), 240.

Aristotelian teaching: Asimov, *Asimov's Biographical Encyclopedia,* 17–20.

61 *retrograde motion:* Katz, *History of Mathematics,* 406.

62 *Aristarchus:* Asimov, *Asimov's Biographical Encyclopedia,* 24–25, and James Evans, "Aristarchus of Samos," *Encyclopedia Britannica,* https://www.britannica.com/biography/Aristarchus-of-Samos.

63 *Archimedes himself realized:* Evans, "Aristarchus of Samos."

Ptolemaic system: Katz, *History of Mathematics,* 145–57.

64 *Giordano Bruno:* Martínez, *Burned Alive.*

Galileo Galilei: The Galileo Project, http://galileo.rice.edu/galileo.html, is an excellent online resource for Galileo's life and work. Fermi and

Bernardini, *Galileo and the Scientific Revolution*, originally published in 1961, is a delightful biography of Galileo for general readers. *Asimov's Biographical Encyclopedia*, 91–96, is a good quick introduction to Galileo, and so is Kline, *Mathematics in Western Culture,* 182–95. For a scholarly treatment, see Drake, *Galileo at Work*, and Michele Camerota, "Galilei, Galileo," in Gillispie, *Complete Dictionary,* 96–103.

Marina Gamba: http://galileo.rice.edu/fam/marina.html.

was his favorite: Sobel, *Galileo's Daughter*. Sister Maria Celeste's letters to her father are at http://galileo.rice.edu/fam/daughter.html#letters.

65 Two New Sciences: The book is available free online at http://oll.libertyfund.org/titles/galilei-dialogues-concerning-two-new-sciences.

66 *proposed that heavy objects fall:* Kline, *Mathematics in Western Culture*, 188–90.

67 *"one-tenth of a pulse-beat":* Galileo, *Discourses*, 179, http://oll.libertyfund.org/titles/753#Galileo_0416_607.

"same ratio as the odd numbers beginning with unity": Ibid., 190, http://oll.libertyfund.org/titles/753#Galileo_0416_516.

69 *"very straight, smooth, and polished":* Ibid., 178, http://oll.libertyfund.org/titles/753#Galileo_0416_607.

70 *"as big as a ship's cable":* Ibid., 109, http://oll.libertyfund.org/titles/753#Galileo_0416_242.

71 *chandelier swaying overhead:* Fermi and Bernardini, *Galileo and the Scientific Revolution*, 17–20, and Kline, *Mathematics in Western Culture*, 182.

72 *"Thousands of times I have observed":* Galileo, *Discourses,* 140, http://oll.libertyfund.org/titles/753#Galileo_0416_338.

"the lengths are to each other as the squares": Ibid., 139, http://oll.libertyfund.org/titles/753#Galileo_0416_335.

73 *"may appear to many exceedingly arid":* Ibid., 138, http://oll.libertyfund.org/titles/753#Galileo_0416_329.

74 *Josephson junction:* Strogatz, *Sync*, chapter 5, and Richard Newrock, "What Are Josephson Junctions? How Do They Work?," *Scientific American*, https://www.scientificamerican.com/article/what-are-josephson-juncti/.

longitude problem: Sobel, *Longitude*.

75 *global positioning system:* Thompson, "Global Positioning System," and https://www.gps.gov.

78 *Johannes Kepler:* For Kepler's life and work, see Owen Gingerich, "Johannes Kepler," in Gillispie, *Complete Dictionary*, vol. 7, online at https://www.encyclopedia.com/people/science-and-technology/astronomy-biographies/johannes-kepler#kjen14, with amendments by J. R. Voelkel in vol. 22. See also Kline, *Mathematics in Western Culture*, 110–25; Edwards, *The Historical Development,* 99–103; Asimov, *Asimov's Biographical*

Encyclopedia, 96–99; Simmons, *Calculus Gems*, 69–83; and Burton, *History of Mathematics*, 355–60.

"criminally inclined": Quoted in Gingerich, "Johannes Kepler," https://www.encyclopedia.com/people/science-and-technology/astronomy-biographies/johannes-kepler#kjen14.

"bad-tempered": Ibid.

"such a superior and magnificent mind": Ibid.

79 *"Day and night I was consumed by the computing":* Ibid.

80 *"God is being celebrated in astronomy":* Ibid.

81 *"this tedious procedure":* Kepler in *Astronomia Nova*, quoted by Owen Gingerich, *The Book Nobody Read: Chasing the Revolutions of Nicolaus Copernicus* (New York: Penguin, 2005), 48.

84 *"sacred frenzy":* Quoted in Gingerich, "Johannes Kepler," https://www.encyclopedia.com/people/science-and-technology/astronomy-biographies/johannes-kepler#kjen14.

85 *"My dear Kepler, I wish we could laugh":* Quoted in Martínez, *Science Secrets*, 34.

86 *"Johannes Kepler became enamored":* Koestler, *The Sleepwalkers*, 33.

4. The Dawn of Differential Calculus

90 *China, India, and the Islamic world:* Katz, "Ideas of Calculus"; Katz, *History of Mathematics,* chapters 6 and 7; and Burton, *History of Mathematics,* 238–85.

91 *Al-Hasan Ibn al-Haytham:* Katz, "Ideas of Calculus," and J. J. O'Connor and E. F. Robertson, "Abu Ali al-Hasan ibn al-Haytham," http://www-history.mcs.st-andrews.ac.uk/Biographies/Al-Haytham.html.

92 *François Viète:* Katz, *History of Mathematics,* 369–75.

decimal fractions: Ibid., 375–78.

93 *Evangelista Torricelli and Bonaventura Cavalieri:* Alexander, *Infinitesimal,* discusses their battles with the Jesuits over infinitesimals, which were seen as dangerous religiously, not just mathematically.

99 *René Descartes:* For his life, see Clarke, *Descartes;* Simmons, *Calculus Gems*, 84–92; and Asimov, *Asimov's Biographical Encyclopedia*, 106–8. For summaries of his math and physics intended for general readers, see Kline, *Mathematics in Western Culture*, 159–81; Edwards, *The Historical Development*; Katz, *History of Mathematics,* sections 11.1 and 12.1; and Burton, *History of Mathematics*, section 8.2. For a scholarly historical treatment of his work in mathematics and physics, see Michael S. Mahoney, "Descartes: Mathematics and Physics," in Gillispie, *Complete Dictionary*, also online at *Encyclopedia Britannica*, https://www.encyclopedia.com

/science/dictionaries-thesauruses-pictures-and-press-releases/descartes-mathematics-and-physics.

"What the ancients have taught us is so scanty": René Descartes, *Les Passions de l'Ame* (1649), quoted in Guicciardini, *Isaac Newton,* 31.

100 *"the country of bears, amid rocks and ice":* Henry Woodhead, *Memoirs of Christina, Queen of Sweden* (London: Hurst and Blackett, 1863), 285.

Pierre de Fermat: Mahoney, *Mathematical Career,* is the definitive treatment. Simmons, *Calculus Gems,* 96–105, is brisk and entertaining about Fermat (just as the author was with everything he wrote; if you haven't read Simmons, you must).

Fermat and Descartes locked horns: Mahoney, *Mathematical Career*, chapter 4.

101 *tried to ruin his reputation:* Ibid., 171.

Fermat came up with them first: I agree with the assessment in Simmons, *Calculus Gems,* 98, about how the credit for analytic geometry should be apportioned: "Superficially Descartes's essay looks as if it might be analytic geometry, but isn't; while Fermat's doesn't look it, but is." For more even-handed views, see Katz, *History of Mathematics,* 432–42, and Edwards, *The Historical Development,* 95–97.

finding a method of analysis: Guicciardini, *Isaac Newton*, and Katz, *History of Mathematics,* 368–69.

102 *"low cunning, deplorable indeed":* Descartes, rule 4 in *Rules for the Direction of the Mind* (1629), as quoted in Katz, *History of Mathematics,* 368–69.

"analysis of the bunglers in mathematics": Quoted in Guicciardini, *Isaac Newton,* 77.

103 *optimization problems:* Mahoney, *Mathematical Career*, 199–201, discusses Fermat's work on the maximization problem considered in the main text.

106 adequality: Ibid., 162–65, and Katz, *History of Mathematics*, 470–72.

107 *JPEG:* Austin, "What Is . . . JPEG?," and Higham et al., *The Princeton Companion,* 813–16.

108 *how day length varies:* Timeanddate.com will give you the information for any location of interest.

112 *sine waves called wavelets:* For a clear introduction to wavelets and their many applications, see Dana Mackenzie, "Wavelets: Seeing the Forest and the Trees," in Beyond Discovery: The Path from Research to Human Benefit, a project of the National Academy of Sciences; go to http://www.nasonline.org/publications/beyond-discovery/wavelets.pdf. Then try Kaiser, *Friendly Guide,* Cipra, "*Parlez-Vous* Wavelets?," or Goriely, *Applied Mathematics,* chapter 6. Daubechies, *Ten Lectures,* was a landmark series of lectures on wavelet mathematics by a pioneer in the field.

Federal Bureau of Investigation used wavelets: Bradley et al., "FBI Wavelet/Scalar Quantization."

113 *mathematicians from the Los Alamos National Lab teamed up with the FBI:* Bradley and Brislawn, "The Wavelet/Scalar Quantization"; Brislawn, "Fingerprints Go Digital"; and https://www.nist.gov/itl/iad/image-group/wsq-bibliography.

115 *Snell's sine law:* Kwan et al., "Who Really Discovered Snell's Law?," and Sabra, *Theories of Light*, 99–105.

116 principle of least time: Mahoney, *Mathematical Career*, 387–402.

117 *"my natural inclination to laziness":* Ibid., 398.

"I can scarcely recover from my astonishment": Ibid., 400 (my translation of Fermat's French).

118 *principle of least action:* Fermat's principle of least time anticipated the more general principle of least action. For entertaining and deeply enlightening discussions of this principle, including its basis in quantum mechanics, see R. P. Feynman, R. B. Leighton, and M. Sands, "The Principle of Least Action," *Feynman Lectures on Physics,* vol. 2, chapter 19 (Reading, MA: Addison-Wesley, 1964), and Feynman, *QED*.

119 *Descartes had his own method:* Katz, *History of Mathematics*, 472–73.

120 *"I have given a general method":* Quoted in Grattan-Guinness, *From the Calculus*, 16.

"I do not even want to name him": Quoted in Mahoney, *Mathematical Career*, 177.

121 *found the area under the curve*: Simmons, *Calculus Gems*, 240–41; and Katz, *History of Mathematics,* 481–84.

his studies still fell short: Katz, *History of Mathematics,* 485, explains why he feels Fermat does not deserve to be considered an inventor of calculus, and he makes a good case.

5. The Crossroads

131 *Logarithms were invented:* Stewart, *In Pursuit of the Unknown*, chapter 2, and Katz, *History of Mathematics,* section 10.4.

137 *paintings allegedly by Vermeer:* Braun, *Differential Equations,* section 1.3.

6. The Vocabulary of Change

159 *Usain Bolt:* Bolt, *Faster than Lightning.*

160 *On that night in Beijing:* Jonathan Snowden, "Remembering Usain Bolt's 100m Gold in 2008," Bleacherreport.com (August 19, 2016), https://bleacherreport.com/articles/2657464-remembering-usain-bolts-100m

-gold-in-2008-the-day-he-became-a-legend, and Eriksen et al., "How Fast." For live video of his astonishing performance, see https://www.youtube.com/watch?v=qslbf8L9nl0 and http://www.nbcolympics.com/video/gold-medal-rewind-usain-bolt-wins-100m-beijing.

"That's just me": Snowden, "Remembering Usain Bolt's."

163 *we want to connect the dots:* My analysis is based on that in A. Oldknow, "Analysing Men's 100m Sprint Times with TI-Nspire," https://rcuksportscience.wikispaces.com/file/view/Analysing+men+100m+Nspire.pdf. The details may differ slightly between the two studies because we used different curve-fitting procedures, but our qualitative conclusions are the same.

165 *researchers were on hand with laser guns:* Graubner and Nixdorf, "Biomechanical Analysis."

166 *"Art," said Picasso:* The quote is from "Picasso Speaks," *The Arts* (May 1923), excerpted in http://www.gallerywalk.org/PM_Picasso.html from Alfred H. Barr Jr., *Picasso: Fifty Years of His Art* (New York: Arno Press, 1980).

7. The Secret Fountain

167 *Isaac Newton:* For biographical information, see Gleick, *Isaac Newton*. See also Westfall, *Never at Rest,* and I. B. Cohen, "Isaac Newton," in vol. 10 of Gillispie, *Complete Dictionary*, with amendments by G. E. Smith and W. Newman in vol. 23. For Newton's mathematics, see Whiteside, *The Mathematical Papers,* vols. 1 and 2; Edwards, *The Historical Development;* Grattan-Guinness, *From the Calculus;* Rickey, "Isaac Newton"; Dunham, *Journey Through Genius;* Katz, *History of Mathematics;* Guicciardini, *Reading the Principia;* Dunham, *The Calculus Gallery;* Simmons, *Calculus Gems;* Guicciardini, *Isaac Newton;* Stillwell, *Mathematics and Its History;* and Burton, *History of Mathematics.*

168 *"between straight and curved lines":* René Descartes, *The Geometry of René Descartes: With a Facsimile of the First Edition*, translated by David E. Smith and Marcia L. Latham (Mineola, NY: Dover, 1954), 91. Within twenty years, Descartes was proved wrong about the impossibility of finding arc lengths exactly for curves; see Katz, *History of Mathematics,* 496–98.

169 *"There is no curved line":* I've updated Newton's spelling here for easier reading. The original was "There is no curve line exprest by any æquation . . . but I can in less then half a quarter of an hower tell whether it may be squared." Letter 193 from Newton to Collins, November 8, 1676,

in Turnbull, *Correspondence of Isaac Newton*, 179. The omitted material involves technical caveats about the class of trinomial equations to which his claim applied. See "A Manuscript by Newton on Quadratures," manuscript 192, in ibid., 178.

"the fountain I draw it from": Letter 193 from Newton to Collins, November 8, 1676, in ibid., 180. Again, I've updated the spelling; Newton wrote "ye fountain."

weren't the first to notice this theorem: Katz, *History of Mathematics*, 498–503, shows that James Gregory and Isaac Barrow had both related the area problem to the tangent problem and so had anticipated the fundamental theorem but concludes that "neither of these men in 1670 could mold these methods into a true computational and problem-solving tool." Five years before that, however, Newton already had. In a sidebar on page 521, Katz makes a convincing case that Newton and Leibniz (as opposed to "Fermat or Barrow or someone else") deserve credit for the invention of calculus.

173 *Scholars in the Middle Ages:* Katz, *History of Mathematics*, section 8.4.

182 *college notebook:* You can explore Newton's handwritten college notebook online. The page shown in the main text is http://cudl.lib.cam.ac.uk/view/MS-ADD-04000/260.

186 *Isaac Newton was born:* My account of Newton's early life is based on Gleick, *Isaac Newton*.

188 *Newton chanced upon something magical:* Whiteside, *The Mathematical Papers*, vol. 1, 96–142, and Katz, *History of Mathematics*, section 12.5. Edwards gives a fascinating treatment of Wallis's work on interpolation and infinite products and shows how Newton's work on power series arose from his attempt to generalize it; see Edwards, *The Historical Development*, chapter 7. We know when Newton made these discoveries because he dated them in an entry on page 14v of his college notebook (online at https://cudl.lib.cam.ac.uk/view/MS-ADD-04000/32). Newton wrote, "I find that in ye year 1664 a little before Christmas I . . . borrowed Wallis' works & by consequence made these Annotations . . . in winter between the years 1664 & 1665. At wch time I found the method of Infinite series. And in summer 1665 being forced from Cambridge by the Plague I computed ye area of ye Hyperbola . . . to two & fifty figures by the same method."

190 *He cooked it up by an argument:* Edwards, *The Historical Development*, 178–87, and Katz, *History of Mathematics*, 506–59, show the steps in Newton's thinking as he derived his results for power series.

192 *"really too much delight in these inventions"*: Letter 188 from Newton

to Oldenburg, October 24, 1676, in Turnbull, *Correspondence of Isaac Newton*, 133.

193 *mathematicians in Kerala, India:* Katz, "Ideas of Calculus"; Katz, *History of Mathematics,* 494–96.

"By their help analysis reaches": This line appears in the famous *epistola prior*, Newton's reply to Leibniz's first inquiry, sent via Henry Oldenburg as intermediary; see letter 165 from Newton to Oldenburg, June 13, 1676, in Turnbull, *Correspondence of Isaac Newton*, 39.

195 *"prime of my age for invention":* Draft letter from Newton to Pierre des Maizeaux, written in 1718, when Newton was seeking to establish his priority over Leibniz in the invention of calculus; available online at https://cudl.lib.cam.ac.uk/view/MS-ADD-03968/1349 in the collection of Cambridge University Library. The full quote is breathtaking: "In the beginning of the year 1665 I found the Method of approximating series & the Rule for reducing any dignity of any Binomial into such a series. The same year in May I found the method of Tangents of Gregory & Slusius, & in November had the direct method of fluxions & the next year in January had the Theory of Colours & in May following I had entrance into y[e] inverse method of fluxions. And the same year I began to think of gravity extending to y[e] orb of the Moon & (having found out how to estimate the force with which a globe revolving within a sphere presses the surface of the sphere) from Kepler's rule of the periodical times of the Planets being in sesquialterate [three-half power] proportion of their distances from the centers of their Orbs, I deduced that the forces which keep the Planets in their Orbs must be reciprocally as the squares of their distances from the centers about which they revolve: & thereby compared the force requisite to keep the Moon in her Orb with the force of gravity at the surface of the earth, & found them answer pretty nearly. All this was in the two plague years of 1665 and 1666. For in those days I was in the prime of my age for invention & minded Mathematicks & Philosophy more than at any time since."

"baited by little smatterers in mathematics": Quoted in Whiteside, "The Mathematical Principles," reference in his ref. 2.

196 *Thomas Hobbes:* Alexander, *Infinitesimal,* tells the story of Hobbes's furious battles with Wallis, which were as political as they were mathematical. Chapter 7 focuses on Hobbes as would-be geometer.

a "scab of symbols": Quoted in Stillwell, *Mathematics and Its History*, 164.

"scurvy book": Ibid.

not "worthy of public utterance": Quoted in Guicciardini, *Isaac Newton*, 343.

"*Our specious algebra":* Ibid.

8. Fictions of the Mind

199 *"His name is Mr. Newton":* Letter from Isaac Barrow to John Collins, August 20, 1669, quoted in Gleick, *Isaac Newton*, 68.

"send me the proof": Letter 158, from Leibniz to Oldenburg, May 2, 1676, in Turnbull, *Correspondence of Isaac Newton*, 4. For more on the Newton-Leibniz correspondence, see Mackinnon, "Newton's Teaser." Guicciardini, *Isaac Newton*, 354–61, offers a particularly clear and helpful analysis of the mathematical cat-and-mouse game taking place between Newton and Leibniz in the letters. The original letters appear in Turnbull, *Correspondence of Isaac Newton;* see especially letters 158 (Leibniz's initial inquiry to Newton via Oldenburg), 165 (Newton's *epistola prior*, terse and intimidating), 172 (Leibniz's request for clarification), 188 (Newton's *epistola posterior*, gentler and clearer but still intended to show Leibniz who was boss), and 209 (Leibniz fighting back, though graciously, and making it clear that he knew calculus too).

200 *"distasteful to me":* One of the best zingers in the *epistola prior*, letter 165 from Newton to Oldenburg, June 13, 1676. See Turnbull, *Correspondence of Isaac Newton*, 39.

"very distinguished": From the *epistola posterior*, letter 188 from Newton to Oldenburg, October 24, 1676, in ibid., 130.

"hope for very great things from him": Ibid.

"the same goal is approached": Ibid.

201 *"I have preferred to conceal it thus":* Ibid., 134. The encryption encodes Newton's understanding of the fundamental theorem and the central problems of calculus: "given any equation involving any number of fluent quantities, to find the fluxions, and conversely." See also page 153, note 25.

"in the twinkling of an eyelid": Letter from Leibniz to Marquis de L'Hospital, 1694, excerpted in Child, *Early Mathematical Manuscripts*, 221. Also quoted in Edwards, *The Historical Development*, 244.

"burdened with a deficiency": Mates, *Philosophy of Leibniz*, 32.

Skinny, stooped, and pale: Ibid.

the most versatile genius: For Leibniz's life, see Hofmann, *Leibniz in Paris;* Asimov, *Asimov's Biographical Encyclopedia;* and Mates, *Philosophy of Leibniz*. For Leibniz's philosophy, see Mates, *Philosophy of Leibniz*. For Leibniz's mathematics, see Child, *Early Mathematical Manuscripts;* Edwards, *The Historical Development;* Grattan-Guinness, *From the Calculus;* Dunham, *Journey Through Genius;* Katz, *History of Mathematics;* Guicciardini, *Reading the Principia;* Dunham, *The Calculus Gallery;* Simmons, *Calculus Gems;* Guicciardini, *Isaac Newton;* Stillwell, *Mathematics and Its History;* and Burton, *History of Mathematics*.

202 *Leibniz's approach to calculus:* Edwards, *The Historical Development*, chapter 9, is especially good. See also Katz, *History of Mathematics*, section 12.6, and Grattan-Guinness, *From the Calculus*, chapter 2.

203 *more pragmatic view:* For example, Leibniz wrote: "We have to make an effort in order to keep pure mathematics chaste from metaphysical controversies. This we will achieve if, without worrying whether the infinites and infinitely smalls in quantities, numbers, and lines are real, we use infinites and infinitely smalls as an appropriate expression for abbreviating reasonings." Quoted in Guicciardini, *Reading the Principia*, 160.

"fictions of the mind": Leibniz in a letter to Des Bosses in 1706, quoted in Guicciardini, *Reading the Principia,* 159.

208 *"My calculus":* Quoted in ibid., 166.

209 *Leibniz deduced the sine law with ease:* Edwards, *The Historical Development,* 259.

"other very learned men": Quoted in ibid.

212 *problem that led him to the fundamental theorem:* Ibid., 236–38. Actually, the sum that concerned Leibniz was the sum of the reciprocals of the triangular numbers, which is twice as large as the sum considered in the main text. See also Grattan-Guinness, *From the Calculus,* 60–62.

218 *"Finding the areas of figures":* From a letter to Ehrenfried Walter von Tschirnhaus in 1679, quoted in Guicciardini, *Reading the Principia*, 145.

the human immunodeficiency virus: For HIV and AIDS statistics, see https://ourworldindata.org/hiv-aids/. For the history of the virus and attempts to combat it, see https://www.avert.org/professionals/history-hiv-aids/overview.

219 *HIV infection typically progressed through three stages:* "The Stages of HIV Infection," AIDSinfo, https://aidsinfo.nih.gov/understanding-hiv-aids/fact-sheets/19/46/the-stages-of-hiv-infection.

220 *Ho and Perelson's work:* Ho et al., "Rapid Turnover"; Perelson et al., "HIV-1 Dynamics"; Perelson, "Modelling Viral and Immune System"; and Murray, *Mathematical Biology 1.*

224 *triple-combination therapy:* The results of the probability calculation first appeared in Perelson et al., "Dynamics of HIV-1."

225 *Man of the Year:* Gorman, "Dr. David Ho."

Perelson received a major prize: American Physical Society, 2017 Max Delbruck Prize in Biological Physics Recipient, https://www.aps.org/programs/honors/prizes/prizerecipient.cfm?first_nm=Alan&last_nm=Perelson&year=2017.

hepatitis C: "Multidisciplinary Team Aids Understanding of Hepatitis C Virus and Possible Cure," Los Alamos National Laboratory, March

2013, http://www.lanl.gov/discover/publications/connections/2013-03/understanding-hep-c.php. For an introduction to the mathematical modeling of hepatitis C, see Perelson and Guedj, "Modelling Hepatitis C."

9. The Logical Universe

228 *Cambrian explosion for mathematics:* For the many offshoots of calculus in the years from 1700 to the present, see Kline, *Mathematics in Western Culture;* Boyer, *The History of the Calculus;* Edwards, *The Historical Development;* Grattan-Guinness, *From the Calculus;* Katz, *History of Mathematics;* Dunham, *The Calculus Gallery;* Stewart, *In Pursuit of the Unknown;* Higham et al., *The Princeton Companion;* and Goriely, *Applied Mathematics.*

229 *system of the world:* Peterson, *Newton's Clock;* Guicciardini, *Reading the Principia;* Stewart, *In Pursuit of the Unknown;* and Stewart, *Calculating the Cosmos.*

ushered in the Enlightenment: Kline, *Mathematics in Western Culture*, 234–86, chronicles the profound impact that Newton's work had on the course of Western philosophy, religion, aesthetics, and literature as well as on science and mathematics. See also W. Bristow, "Enlightenment," https://plato.stanford.edu/entries/enlightenment/.

"made his head ache": D. Brewster, *Memoirs of the Life, Writings, and Discoveries of Sir Isaac Newton*, vol. 2 (Edinburgh: Thomas Constable, 1855), 158.

232 *when an apple fell:* For the surprising history of the apple story, see Gleick, *Isaac Newton*, 55–57, and note 18 on 207. See also Martínez, *Science Secrets*, chapter 3.

233 *"force requisite to keep the Moon in her Orb":* Draft letter from Newton to Pierre des Maizeaux, written in 1718, available online at https://cudl.lib.cam.ac.uk/view/MS-ADD-03968/1349 in the collection of Cambridge University Library.

234 *"In ellipses":* Asimov, *Asimov's Biographical Encyclopedia*, 138, gives one version of this oft-told story.

followed as logical necessities: Katz, *History of Mathematics*, 516–19, outlines Newton's geometric arguments. Guicciardini, *Reading the Principia*, discusses how Newton's contemporaries reacted to the *Principia* and what their criticisms of it were (some of their objections were cogent). A modern derivation of Kepler's laws from the inverse-square law is given by Simmons, *Calculus Gems*, 326–35.

237 *Neptune:* Jones, *John Couch Adams,* and Sheehan and Thurber, "John Couch Adams's Asperger Syndrome."

Katherine Johnson: Shetterly, *Hidden Figures,* gave Katherine Johnson the recognition she so long deserved. For more about her life, see https://www.nasa.gov/content/katherine-johnson-biography. For her mathematics, see Skopinski and Johnson, "Determination of Azimuth Angle." See also http://www-groups.dcs.st-and.ac.uk/history/Biographies/Johnson_Katherine.html and https://ima.org.uk/5580/hidden-figures-impact-mathematics/.

238 *NASA official reminded the audience:* Sarah Lewin, "NASA Facility Dedicated to Mathematician Katherine Johnson," Space.com, May 5, 2016, https://www.space.com/32805-katherine-johnson-langley-building-dedication.html.

239 *boisterous toast:* Quoted in Kline, *Mathematics in Western Culture,* 282. The account of the dinner party comes from the diary of the party's host, the painter Benjamin Haydon, excerpted in Ainger, *Charles Lamb,* 84–86.

Thomas Jefferson: Cohen, *Science and the Founding Fathers,* makes a persuasive case for Newton's influence on Jefferson and the "Newtonian echoes" in the Declaration of Independence; also see "The Declaration of Independence," http://math.virginia.edu/history/Jefferson/jeff_r(4).htm. For more on Jefferson and mathematics, see the lecture by John Fauvel, "'When I Was Young, Mathematics Was the Passion of My Life': Mathematics and Passion in the Life of Thomas Jefferson," online at http://math.virginia.edu/history/Jefferson/jeff_r.htm.

240 *"I have given up newspapers":* Letter from Thomas Jefferson to John Adams, January 21, 1812, online at https://founders.archives.gov/documents/Jefferson/03-04-02-0334.

moldboard of a plow: Cohen, *Science and the Founding Fathers,* 101. See also "Moldboard Plow," *Thomas Jefferson Encyclopedia*, https://www.monticello.org/site/plantation-and-slavery/moldboard-plow, and "Dig Deeper—Agricultural Innovations," https://www.monticello.org/site/jefferson/dig-deeper-agricultural-innovations.

"what it promises in theory": Letter from Thomas Jefferson to Sir John Sinclair, March 23, 1798, https://founders.archives.gov/documents/Jefferson/01-30-02-0135.

241 *"Unless I am much mistaken":* Hall and Hall, *Unpublished Scientific Papers,* 281.

242 ordinary differential equations: For ordinary differential equations and their applications, see Simmons, *Differential Equations.* See also Braun,

Differential Equations; Strogatz, *Nonlinear Dynamics;* Higham et al., *The Princeton Companion;* and Goriely, *Applied Mathematics.*

244 partial differential equation: For partial differential equations and their applications, see Farlow, *Partial Differential Equations*, and Haberman, *Applied Partial Differential Equations*. See also Higham et al., *The Princeton Companion,* and Goriely, *Applied Mathematics.*

245 *Boeing 787 Dreamliner:* Norris and Wagner, *Boeing 787,* and http://www.boeing.com/commercial/787/by-design/#/featured.

246 *aeroelastic flutter:* Jason Paur, "Why 'Flutter' Is a 4-Letter Word for Pilots," *Wired* (March 25, 2010), https://www.wired.com/2010/03/flutter-testing-aircraft/.

247 *Black-Scholes model for pricing financial options:* Szpiro, *Pricing the Future,* and Stewart, *In Pursuit of the Unknown*, chapter 17.

248 *Hodgkin-Huxley model:* Ermentrout and Terman, *Mathematical Foundations*, and Rinzel, "Discussion."

Einstein's general theory of relativity: Stewart, *In Pursuit of the Unknown*, chapter 13, and Ferreira, *Perfect Theory*. See also Greene, *The Elegant Universe*, and Isaacson, *Einstein.*

Schrödinger equation: Stewart, *In Pursuit of the Unknown*, chapter 14.

10. Making Waves

249 *Fourier:* Körner, *Fourier Analysis,* and Kline, *Mathematics in Western Culture*, chapter 19. For his life and work, see Dirk J. Struik, "Joseph Fourier," *Encyclopedia Britannica*, https://www.britannica.com/biography/Joseph-Baron-Fourier. See also Grattan-Guinness, *From the Calculus;* Stewart, *In Pursuit of the Unknown;* Higham et al., *The Princeton Companion;* and Goriely, *Applied Mathematics.*

heat flow: The mathematics of Fourier's heat equation is discussed in Farlow, *Partial Differential Equations*, Katz, *History of Mathematics*, and Haberman, *Applied Partial Differential Equations.*

252 *wave equation:* For the mathematics of vibrating strings, Fourier series, and the wave equation, see Farlow, *Partial Differential Equations*; Katz, *History of Mathematics;* Haberman, *Applied Partial Differential Equations;* Stillwell, *Mathematics and Its History;* Burton, *History of Mathematics;* Stewart, *In Pursuit of the Unknown;* and Higham et al., *The Princeton Companion.*

259 *Chladni patterns:* The original images are reproduced at https://publicdomainreview.org/collections/chladni-figures-1787/ and http://www.sites.hps.cam.ac.uk/whipple/explore/acoustics/ernstchladni

/chladniplates/. For a modern demo, see the video by Steve Mould called "Random Couscous Snaps into Beautiful Patterns," https://www.youtube.com/watch?v=CR_XL192wXw&feature=youtu.be and the video by Physics Girl called "Singing Plates—Standing Waves on Chladni Plates," https://www.youtube.com/watch?v=wYoxOJDrZzw.

261 *Sophie Germain:* Her theory of Chladni patterns is discussed in Bucciarelli and Dworsky, *Sophie Germain*. For biographies, see: https://www.agnesscott.edu/lriddle/women/germain.htm and http://www.pbs.org/wgbh/nova/physics/sophie-germain.html and http://www-groups.dcs.st-and.ac.uk/~history/Biographies/Germain.html.

262 "*the noblest courage":* Quoted in Newman, *The World of Mathematics*, vol. 1, 333.

microwave oven: For a very clear explanation of how a microwave oven works as well as a demonstration of the experiment I suggested, see "How a Microwave Oven Works," https://www.youtube.com/watch?v=kp33ZprO0Ck. To measure the speed of light with a microwave oven, you can also use chocolate, as shown here: https://www.youtube.com/watch?v=GH5W6xEeY5U. For the backstory of microwave ovens and the gooey, sticky mess that Percy Spencer felt in his pocket, see Matt Blitz, "The Amazing True Story of How the Microwave Was Invented by Accident," *Popular Mechanics* (February 23, 2016), https://www.popularmechanics.com/technology/gadgets/a19567/how-the-microwave-was-invented-by-accident/.

265 *CT scanning:* Kevles, *Naked to the Bone,* 145–72; Goriely, *Applied Mathematics*, 85–89; and https://www.nobelprize.org/nobel_prizes/medicine/laureates/1979/. The original paper that solves the reconstruction problem with calculus and Fourier series is Cormack, "Representation of a Function."

267 *Allan Cormack:* The original paper that solves the reconstruction problem for computerized tomography by using calculus, Fourier series, and integral equations is Cormack, "Representation of a Function." His Nobel Prize lecture is available online at https://www.nobelprize.org/nobel_prizes/medicine/laureates/1979/cormack-lecture.pdf.

268 *the Beatles:* For the story of Godfrey Hounsfield, the Beatles, and the invention of the CT scanner, see Goodman, "The Beatles," and https://www.nobelprize.org/nobel_prizes/medicine/laureates/1979/perspectives.html.

269 *Cormack explained:* The quote appears on page 563 of his Nobel lecture: https://www.nobelprize.org/nobel_prizes/medicine/laureates/1979/cormack-lecture.pdf.

11. The Future of Calculus

275 writhing number: Fuller, "The Writhing Number." See also Pohl, "DNA and Differential Geometry."

geometry and topology of DNA: Bates and Maxwell, *DNA Topology,* and Wasserman and Cozzarelli, "Biochemical Topology."

knot theory and tangle calculus: Ernst and Sumners, "Calculus for Rational Tangles."

276 *targets for cancer-chemotherapy drugs:* Liu, "DNA Topoisomerase Poisons."

277 *Pierre Simon Laplace:* Kline, *Mathematics in Western Culture;* C. Hoefer, "Causal Determinism," https://plato.stanford.edu/entries/determinism-causal/.

"nothing would be uncertain": Laplace, *Philosophical Essay on Probabilities,* 4.

Sofia Kovalevskaya: Cooke, *Mathematics of Sonya Kovalevskaya,* and Goriely, *Applied Mathematics*, 54–57. She is often referred to by other names; Sonia Kovalevsky is a common variant. For online biographies, see Becky Wilson, "Sofia Kovalevskaya," *Biographies of Women Mathematicians,* https://www.agnesscott.edu/lriddle/women/kova.htm, and J. J. O'Connor and E. F. Robertson, "Sofia Vasilyevna Kovalevskaya," http://www-groups.dcs.st-and.ac.uk/history/Biographies/Kovalevskaya.html.

278 *chaotic tumbling of Hyperion:* Wisdom et al., "Chaotic Rotation."

281 *Poincaré thought he'd solved it:* Diacu and Holmes, *Celestial Encounters.*

Chaotic systems: Gleick, *Chaos;* Stewart, *Does God Play Dice?;* and Strogatz, *Nonlinear Dynamics.*

predictability horizon: Lighthill, "The Recently Recognized Failure."

horizon of predictability for the entire solar system: Sussman and Wisdom, "Chaotic Evolution."

282 *Poincaré's Visual Approach:* Gleick, *Chaos;* Stewart, *Does God Play Dice?;* Strogatz, *Nonlinear Dynamics;* and Diacu and Holmes, *Celestial Encounters.*

284 *Mary Cartwright:* McMurran and Tattersall, "Mathematical Collaboration," and L. Jardine, "Mary, Queen of Maths," *BBC News Magazine*, https://www.bbc.com/news/magazine-21713163. For biographies, see http://www.ams.org/notices/199902/mem-cartwright.pdf and http://www-history.mcs.st-and.ac.uk/Biographies/Cartwright.html.

"very objectionable-looking differential equations": Quoted in L. Jardine, "Mary, Queen of Maths."

285 *"equation itself was to blame":* Dyson, "Review of *Nature's Numbers.*"

287 *Hodgkin and Huxley:* Ermentrout and Terman, *Mathematical Foundations;* Rinzel, "Discussion"; and Edelstein-Keshet, *Mathematical Models.*

Mathematical biology: For introductions to the mathematical modeling of epidemics, heart rhythms, cancer, and brain tumors, see Edelstein-Keshet, *Mathematical Models;* Murray, *Mathematical Biology 1;* and Murray, *Mathematical Biology 2.*

290 complex systems: Mitchell, *Complexity.*

291 *computer chess:* For background on AlphaZero and computer chess, see https://www.technologyreview.com/s/609736/alpha-zeros-alien-chess-shows-the-power-and-the-peculiarity-of-ai/. The original preprint describing AlphaZero is at https://arxiv.org/abs/1712.01815. For video analyses of the games between AlphaZero and Stockfish, start with https://www.youtube.com/watch?v=Ud8F-cNsa-k and https://www.youtube.com/watch?v=6z1o48Sgrck.

293 *the dusk of insight:* Davies, "Whither Mathematics?," https://www.ams.org/notices/200511/comm-davies.pdf.

294 *Paul Erdős:* Hoffman, *The Man Who Loved Only Numbers.*

Conclusion

296 *quantum electrodynamics:* Feynman, *QED*, and Farmelo, *The Strangest Man.*

the most accurate theory: Peskin and Schroeder, *Introduction to Quantum Field Theory,* 196–98. For background, see http://scienceblogs.com/principles/2011/05/05/the-most-precisely-tested-theo/.

297 *Paul Dirac:* For Dirac's life and work, see Farmelo, *The Strangest Man*. The 1928 paper that introduced the Dirac equation is Dirac, "The Quantum Theory."

298 *In 1931 he published a paper:* Dirac, "Quantised Singularities."

"one would be surprised": Ibid., 71.

PET scans: Kevles, *Naked to the Bone*, 201–27, and Higham et al., *The Princeton Companion,* 816–23. For positrons in PET scanning, see Farmelo, *The Strangest Man*, and Rich, "Brief History."

299 *Albert Einstein:* Isaacson, *Einstein*, and Pais, *Subtle Is the Lord.*

general relativity: Ferreira, *Perfect Theory*, and Greene, *The Elegant Universe.*

strange effect on time: For more on GPS and relativistic effects on timekeeping, see Stewart, *In Pursuit of the Unknown,* and http://www.astronomy.ohio-state.edu/~pogge/Ast162/Unit5/gps.html.

300 *gravitational waves:* Levin, *Black Hole Blues*, is a lyrical book about the

search for gravitational waves. For more background, see https://brilliant.org/wiki/gravitational-waves/ and https://www.nobelprize.org/nobel_prizes/physics/laureates/2017/press.html. For the role of calculus, computers, and numerical methods in the discovery, see R. A. Eisenstein, "Numerical Relativity and the Discovery of Gravitational Waves," https://arxiv.org/pdf/1804.07415.pdf.

Bibliography

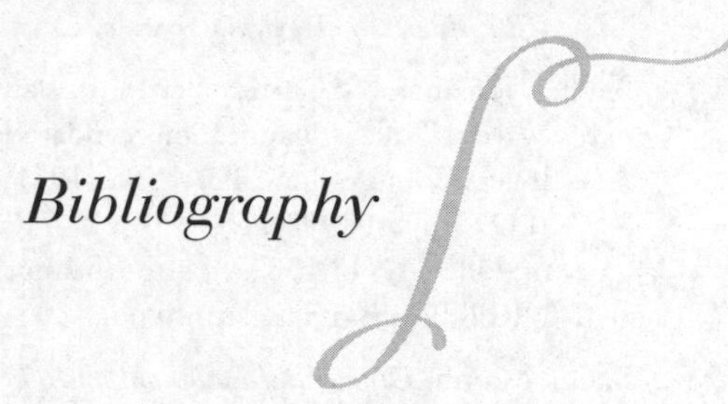

Adams, Douglas. *The Hitchhiker's Guide to the Galaxy.* London: Pan Books, 1979.

Ainger, Alfred. *Charles Lamb.* New York: Harper and Brothers, 1882.

Alexander, Amir. *Infinitesimal: How a Dangerous Mathematical Theory Shaped the Modern World.* New York: Farrar, Straus and Giroux, 2014.

Asimov, Isaac. *Asimov's Biographical Encyclopedia of Science and Technology*. Rev. ed. New York: Doubleday, 1972.

Austin, David. "What Is . . . JPEG?," *Notices of the American Mathematical Society* 55, no. 2 (2008): 226–29. http://www.ams.org/notices/200802/tx080200226p.pdf.

Ball, Philip. "A Century Ago Einstein Sparked the Notion of the Laser." *Physics World* (August 31, 2017). https://physicsworld.com/a/a-century-ago-einstein-sparked-the-notion-of-the-laser/.

Barrow, John D., and Frank J. Tipler. *The Anthropic Cosmological Principle.* New York: Oxford University Press, 1986.

Bates, Andrew D., and Anthony Maxwell. *DNA Topology.* New York: Oxford University Press, 2005.

Bolt, Usain. *Faster than Lightning: My Autobiography.* New York: HarperSport, 2013.

Boyer, Carl B. *The History of the Calculus and Its Conceptual Development.* Mineola, NY: Dover, 1959.

Bradley, Jonathan N., and Christopher M. Brislawn. "The Wavelet/Scalar

Quantization Compression Standard for Digital Fingerprint Images." *IEEE International Symposium on Circuits and Systems* 3 (1994): 205–8.

Bradley, Jonathan N., Christopher M. Brislawn, and Thomas Hopper. "FBI Wavelet/Scalar Quantization Standard for Gray-Scale Fingerprint Image Compression." Proc. SPIE 1961, Visual Information Processing II (27 August 1993). DOI: 10.1117/12.150973; https://doi.org/10.1117/12.150973; http://helmut.knaust.info/class/201330_NREUP/spie93_Fingerprint.pdf.

Braun, Martin. *Differential Equations and Their Applications*. 3rd ed. New York: Springer, 1983.

Brislawn, Christopher M. "Fingerprints Go Digital." *Notices of the American Mathematical Society* 42, no. 11 (1995): 1278–83.

Bucciarelli, Louis L., and Nancy Dworsky. *Sophie Germain: An Essay in the History of Elasticity.* Dordrecht, Netherlands: D. Reidel, 1980.

Burkert, Walter. *Lore and Science in Ancient Pythagoreanism*. Translated by E. L. Minar Jr. Cambridge, MA: Harvard University Press, 1972.

Burton, David M. *The History of Mathematics*. 7th ed. New York: McGraw-Hill, 2011.

Calaprice, Alice. *The Ultimate Quotable Einstein.* Princeton, NJ: Princeton University Press, 2011.

Carroll, Sean. *The Big Picture: On the Origins of Life, Meaning, and the Universe Itself.* New York: Dutton, 2016.

Child, J. M. *The Early Mathematical Manuscripts of Leibniz.* Chicago: Open Court, 1920.

Cipra, Barry. "*Parlez-Vous* Wavelets?" *What's Happening in the Mathematical Sciences* 2 (1994): 23–26.

Clarke, Desmond. *Descartes: A Biography.* Cambridge: Cambridge University Press, 2006.

Cohen, I. Bernard. *Science and the Founding Fathers: Science in the Political Thought of Thomas Jefferson, Benjamin Franklin, John Adams, and James Madison.* New York: W. W. Norton, 1995.

Cooke, Roger. *The Mathematics of Sonya Kovalevskaya.* New York: Springer, 1984.

Cormack, Allan M. "Representation of a Function by Its Line Integrals, with Some Radiological Applications." *Journal of Applied Physics* 34, no. 9 (1963): 2722–27.

Daubechies, Ingrid C. *Ten Lectures on Wavelets.* Philadelphia: Society for Industrial and Applied Mathematics, 1992.

Davies, Brian. "Whither Mathematics?" *Notices of the American Mathematical Society* 52, no. 11 (2005): 1350–56.

Davies, Paul. *The Goldilocks Enigma: Why Is the Universe Just Right for Life?* London: Allen Lane, 2006.

DeRose, Tony, Michael Kass, and Tien Truong. "Subdivision Surfaces in Character Animation." *Proceedings of the 25th Annual Conference on Computer Graphics and Interactive Techniques* (1998): 85–94. DOI: http://dx.doi.org/10.1145/280814.280826; https://graphics.pixar.com/library/Geri/paper.pdf.

Deuflhard, Peter, Martin Weiser, and Stefan Zachow. "Mathematics in Facial Surgery." *Notices of the American Mathematical Society* 53, no. 9 (2006): 1012–16.

Diacu, Florin, and Philip Holmes. *Celestial Encounters: The Origins of Chaos and Stability.* Princeton, NJ: Princeton University Press, 1996.

Dijksterhuis, Eduard J. *Archimedes.* Princeton, NJ: Princeton University Press, 1987.

Dirac, Paul A. M. "Quantised Singularities in the Electromagnetic Field." *Proceedings of the Royal Society of London A* 133 (1931): 60–72. DOI: 10.1098/rspa.1931.0130.

———. "The Quantum Theory of the Electron." *Proceedings of the Royal Society of London A* 117 (1928): 610–24. DOI: 10.1098/rspa.1928.0023.

Drake, Stillman. *Galileo at Work: His Scientific Biography.* Chicago: University of Chicago Press, 1978.

Dunham, William. *The Calculus Gallery: Masterpieces from Newton to Lebesgue.* Princeton, NJ: Princeton University Press, 2005.

———. *Journey Through Genius.* New York: John Wiley and Sons, 1990.

Dyson, Freeman J. "Review of *Nature's Numbers* by Ian Stewart." *American Mathematical Monthly* 103, no. 7 (August/September 1996): 610–12. DOI: 10.2307/2974684.

Edelstein-Keshet, Leah. *Mathematical Models in Biology.* 8th ed. Philadelphia: Society for Industrial and Applied Mathematics, 2005.

Edwards, C. H., Jr. *The Historical Development of the Calculus.* New York: Springer, 1979.

Einstein, Albert. "Physics and Reality." *Journal of the Franklin Institute* 221, no. 3 (1936): 349–82.

———. "Zur Quantentheorie der Strahlung (On the Quantum Theory of Radiation)." *Physikalische Zeitschrift* 18 (1917): 121–28. English translation at http://web.ihep.su/dbserv/compas/src/einstein17/eng.pdf.

Eriksen, H. K., J. R. Kristiansen, Ø. Langangen, and I. K. Wehus. "How Fast Could Usain Bolt Have Run? A Dynamical Study." *American Journal of Physics* 77, no. 3 (2009): 224–28.

Ermentrout, G. Bard, and David H. Terman. *Mathematical Foundations of Neuroscience*. New York: Springer, 2010.

Ernst, Claus, and DeWitt Sumners. "A Calculus for Rational Tangles: Applications to DNA Recombination." *Mathematical Proceedings of the Cambridge Philosophical Society* 108, no. 3 (1990): 489–515.

Farlow, Stanley J. *Partial Differential Equations for Scientists and Engineers.* Mineola, NY: Dover, 1993.

Farmelo, Graham. *The Strangest Man: The Hidden Life of Paul Dirac, Mystic of the Atom.* New York: Basic Books, 2009.

Fermi, Laura, and Gilberto Bernardini. *Galileo and the Scientific Revolution.* Mineola, NY: Dover, 2003.

Ferreira, Pedro G. *The Perfect Theory.* Boston: Houghton Mifflin Harcourt, 2014.

Feynman, Richard P. *QED: The Strange Theory of Light and Matter.* Princeton, NJ: Princeton University Press, 1986.

Forbes, Nancy, and Basil Mahon. *Faraday, Maxwell, and the Electromagnetic Field: How Two Men Revolutionized Physics.* New York: Prometheus Books, 2014.

Fuller, F. Brock. "The Writhing Number of a Space Curve." *Proceedings of the National Academy of Sciences* 68, no. 4 (1971): 815–19.

Galilei, Galileo. *Discourses and Mathematical Demonstrations Concerning Two New Sciences* (1638). Translated from the Italian and Latin into English by Henry Crew and Alfonso de Salvio, with an introduction by Antonio Favaro. New York: Macmillan, 1914. http://oll.libertyfund.org/titles/753.

Gill, Peter. *42: Douglas Adams' Amazingly Accurate Answer to Life, the Universe and Everything.* London: Beautiful Books, 2011.

Gillispie, Charles C., ed. *Complete Dictionary of Scientific Biography.* 26 vols. New York: Charles Scribner's Sons, 2008. Available electronically through the Gale Virtual Reference Library.

Gleick, James. *Chaos: Making a New Science.* New York: Viking, 1987.

———. *Isaac Newton.* New York: Pantheon, 2003.

Goodman, Lawrence R. "The Beatles, the Nobel Prize, and CT Scanning of the Chest." *Thoracic Surgery Clinics* 20, no. 1 (2010): 1–7. https://www.thoracic.theclinics.com/article/S1547–4127(09)00090–5/fulltext. DOI: https://doi.org/10.1016/j.thorsurg.2009.12.001.

Goriely, Alain. *Applied Mathematics: A Very Short Introduction.* Oxford: Oxford University Press, 2018.

Gorman, Christine. "Dr. David Ho: The Disease Detective." *Time* (December 30, 1996). http://content.time.com/time/magazine/article/0,9171,135255,00.html.

Grattan-Guinness, Ivor, ed. *From the Calculus to Set Theory, 1630–1910: An Introductory History.* Princeton, NJ: Princeton University Press, 1980.

Graubner, Rolf, and Eberhard Nixdorf. "Biomechanical Analysis of the Sprint and Hurdles Events at the 2009 IAAF World Championships in Athletics." *New Studies in Athletics* 26, nos. 1/2 (2011): 19–53.

Greene, Brian. *The Elegant Universe: Superstrings, Hidden Dimensions and the Quest for the Ultimate Theory.* New York: W. W. Norton, 1999.

Guicciardini, Niccolò. *Isaac Newton on Mathematical Certainty and Method.* Cambridge, MA: MIT Press, 2009.

———. *Reading the Principia: The Debate on Newton's Mathematical Methods for Natural Philosophy from 1687 to 1736.* Cambridge: Cambridge University Press, 1999.

Guthrie, Kenneth S. *The Pythagorean Sourcebook and Library.* Grand Rapids, MI: Phanes Press, 1987.

Haberman, Richard. *Applied Partial Differential Equations.* 4th ed. Upper Saddle River, NJ: Prentice Hall, 2003.

Hall, A. Rupert, and Marie Boas Hall, eds. *Unpublished Scientific Papers of Isaac Newton.* Cambridge: Cambridge University Press, 1962.

Hamming, Richard W. "The Unreasonable Effectiveness of Mathematics." *American Mathematical Monthly* 87, no. 2 (1980): 81–90. https://www.dartmouth.edu/~matc/MathDrama/reading/Hamming.html.

Heath, Thomas L., ed. *The Works of Archimedes.* Mineola, NY: Dover, 2002.

Higham, Nicholas J., Mark R. Dennis, Paul Glendinning, Paul A. Martin, Fadil Santosa, and Jared Tanner, eds. *The Princeton Companion to Applied Mathematics.* Princeton, NJ: Princeton University Press, 2015.

Ho, David D., Avidan U. Neumann, Alan S. Perelson, Wen Chen, John M. Leonard, and Martin Markowitz. "Rapid Turnover of Plasma Virions and CD4 Lymphocytes in HIV-1 Infection." *Nature* 373, no. 6510 (1995): 123–26.

Hoffman, Paul. *The Man Who Loved Only Numbers: The Story of Paul Erdős and the Search for Mathematical Truth.* New York: Hachette, 1998.

Hofmann, Joseph E. *Leibniz in Paris 1672–1676: His Growth to Mathematical Maturity.* Cambridge: Cambridge University Press, 1972.

Isaacson, Walter. *Einstein: His Life and Universe.* New York: Simon and Schuster, 2007.

Isacoff, Stuart. *Temperament: How Music Became a Battleground for the Great Minds of Western Civilization.* New York: Knopf, 2001.

Jones, H. S. *John Couch Adams and the Discovery of Neptune.* Cambridge: Cambridge University Press, 1947.

Kaiser, Gerald. *A Friendly Guide to Wavelets.* Boston: Birkhäuser, 1994.

Katz, Victor J. *A History of Mathematics: An Introduction*. 2nd ed. Boston: Addison Wesley Longman, 1998.

———. "Ideas of Calculus in Islam and India." *Mathematics Magazine* 68, no. 3 (1995): 163–74.

Kevles, Bettyann H. *Naked to the Bone: Medical Imaging in the Twentieth Century.* Rutgers, NJ: Rutgers University Press, 1997.

Kline, Morris. *Mathematics in Western Culture.* London: Oxford University Press, 1953.

Koestler, Arthur. *The Sleepwalkers: A History of Man's Changing Vision of the Universe.* New York: Penguin, 1990.

Körner, Thomas W. *Fourier Analysis.* Cambridge: Cambridge University Press, 1989.

Kwan, Alistair, John Dudley, and Eric Lantz. "Who Really Discovered Snell's Law?" *Physics World* 15, no. 4 (2002): 64.

Laplace, Pierre Simon. *A Philosophical Essay on Probabilities.* Translated by

Frederick Wilson Truscott and Frederick Lincoln Emory. New York: John Wiley and Sons, 1902.

Laubenbacher, Reinhard, and David Pengelley. *Mathematical Expeditions: Chronicles by the Explorers.* New York: Springer, 1999.

Levin, Janna. *Black Hole Blues and Other Songs from Outer Space.* New York: Knopf, 2016.

Lighthill, James. "The Recently Recognized Failure of Predictability in Newtonian Dynamics." *Proceedings of the Royal Society of London A* 407, no. 1832 (1986): 35–50.

Liu, Leroy F. "DNA Topoisomerase Poisons as Antitumor Drugs." *Annual Review of Biochemistry* 58, no. 1 (1989): 351–75.

Livio, Mario. *Is God a Mathematician?* New York: Simon and Schuster, 2009.

Mackinnon, Nick. "Newton's Teaser." *Mathematical Gazette* 76, no. 475 (1992): 2–27.

Mahoney, Michael S. *The Mathematical Career of Pierre de Fermat 1601–1665.* 2nd ed. Princeton, NJ: Princeton University Press, 1994.

Martínez, Alberto A. *Burned Alive: Giordano Bruno, Galileo and the Inquisition.* London: Reaktion Books, 2018.

———. *The Cult of Pythagoras: Math and Myths.* Pittsburgh: University of Pittsburgh Press, 2012.

———. *Science Secrets: The Truth About Darwin's Finches, Einstein's Wife, and Other Myths.* Pittsburgh: University of Pittsburgh Press, 2011.

Mates, Benson. *The Philosophy of Leibniz: Metaphysics and Language.* Oxford: Oxford University Press, 1986.

Maxwell, James Clerk. "On Physical Lines of Force. Part III. The Theory of Molecular Vortices Applied to Statical Electricity." *Philosophical Magazine* (April/May 1861): 12–24.

Mazur, Joseph. *Zeno's Paradox: Unraveling the Ancient Mystery Behind the Science of Space and Time.* New York: Plume, 2008.

McAdams, Aleka, Stanley Osher, and Joseph Teran. "Crashing Waves, Awesome Explosions, Turbulent Smoke, and Beyond: Applied Mathematics and Scientific Computing in the Visual Effects Industry." *Notices of the American Mathematical Society* 57, no. 5 (2010): 614–23. https://www.ams.org/notices/201005/rtx100500614p.pdf.

McMurran, Shawnee L., and James J. Tattersall. "The Mathematical Collaboration of M. L. Cartwright and J. E. Littlewood." *American Mathematical Monthly* 103, no. 10 (December 1996): 833–45. DOI: 10.2307/2974608.

Mitchell, Melanie. *Complexity: A Guided Tour.* Oxford: Oxford University Press, 2011.

Murray, James D. *Mathematical Biology 1.* 3rd ed. New York: Springer, 2007.

———. *Mathematical Biology 2.* 3rd ed. New York: Springer, 2011.

Netz, Reviel, and William Noel. *The Archimedes Codex: How a Medieval Prayer Book Is Revealing the True Genius of Antiquity's Greatest Scientist.* Boston: Da Capo Press, 2007.

Newman, James R. *The World of Mathematics.* 4 vols. New York: Simon and Schuster, 1956.

Norris, Guy, and Mark Wagner. *Boeing 787 Dreamliner.* Minneapolis: Zenith Press, 2009.

Pais, A. *Subtle Is the Lord.* Oxford: Oxford University Press, 1982.

Perelson, Alan S. "Modelling Viral and Immune System Dynamics." *Nature Reviews Immunology* 2, no. 1 (2002): 28–36.

Perelson, Alan S., Paulina Essunger, and David D. Ho. "Dynamics of HIV-1 and CD4+ Lymphocytes in Vivo." *AIDS* 11, supplement A (1997): S17–S24.

Perelson, Alan S., and Jeremie Guedj. "Modelling Hepatitis C Therapy—Predicting Effects of Treatment." *Nature Reviews Gastroenterology and Hepatology* 12, no. 8 (2015): 437–45.

Perelson, Alan S., Avidan U. Neumann, Martin Markowitz, John M. Leonard, and David D. Ho. "HIV-1 Dynamics in Vivo: Virion Clearance Rate, Infected Cell Life-Span, and Viral Generation Time." *Science* 271, no. 5255 (1996): 1582–86.

Peskin, Michael E., and Daniel V. Schroeder. *An Introduction to Quantum Field Theory.* Boulder, CO: Westview Press, 1995.

Peterson, Ivars. *Newton's Clock: Chaos in the Solar System.* New York: W. H. Freeman, 1993.

Pohl, William F. "DNA and Differential Geometry." *Mathematical Intelligencer* 3, no. 1 (1980): 20–27.

Purcell, Edward M. *Electricity and Magnetism*. 2nd ed. Cambridge: Cambridge University Press, 2011.

Rees, Martin. *Just Six Numbers: The Deep Forces That Shape the Universe.* New York: Basic Books, 2001.

Rich, Dayton A. "A Brief History of Positron Emission Tomography." *Journal of Nuclear Medicine Technology* 25 (1997): 4–11. http://tech.snmjournals.org/content/25/1/4.full.pdf.

Rickey, V. Frederick. "Isaac Newton: Man, Myth, and Mathematics." *College Mathematics Journal* 18, no. 5 (1987): 362–89.

Rinzel, John. "Discussion: Electrical Excitability of Cells, Theory and Experiment: Review of the Hodgkin-Huxley Foundation and an Update." *Bulletin of Mathematical Biology* 52, nos. 1/2 (1990): 5–23.

Robinson, Andrew. "Einstein Said That—Didn't He?" *Nature* 557 (2018): 30. https://www.nature.com/articles/d41586-018-05004-4.

Rorres, Chris, ed. *Archimedes in the Twenty-First Century.* Boston: Birkhäuser, 2017.

Sabra, A. I. *Theories of Light: From Descartes to Newton.* Cambridge: Cambridge University Press, 1981.

Schaffer, Simon. "The Laird of Physics." *Nature* 471 (2011): 289–91.

Schrödinger, Erwin. *Science and Humanism.* Cambridge: Cambridge University Press, 1951.

Sheehan, William, and Steven Thurber. "John Couch Adams's Asperger Syndrome and the British Non-Discovery of Neptune." *Notes and Records* 61, no. 3 (2007): 285–99. http://rsnr.royalsocietypublishing.org/content/61/3/285. DOI: 10.1098/rsnr.2007.0187.

Shetterly, Margot Lee. *Hidden Figures: The American Dream and the Untold Story of the Black Women Mathematicians Who Helped Win the Space Race.* New York: William Morrow, 2016.

Simmons, George F. *Calculus Gems: Brief Lives and Memorable Mathematics.* Washington, DC: Mathematical Association of America, 2007.

———. *Differential Equations with Applications and Historical Notes.* 3rd ed. Boca Raton, FL: CRC Press, 2016.

Skopinski, Ted H., and Katherine G. Johnson. "Determination of Azimuth Angle at Burnout for Placing a Satellite Over a Selected Earth Position."

NASA Technical Report, NASA-TN-D-233, L-289 (1960). https://ntrs.nasa.gov/archive/nasa/casi.ntrs.nasa.gov/19980227091.pdf.

Sobel, Dava. *Galileo's Daughter: A Historical Memoir of Science, Faith, and Love.* New York: Walker, 1999.

———. *Longitude: The True Story of a Lone Genius Who Solved the Greatest Scientific Problem of His Time.* New York: Walker, 1995.

Stein, Sherman. *Archimedes: What Did He Do Besides Cry Eureka?* Washington, DC: Mathematical Association of America, 1999.

Stewart, Ian. *Calculating the Cosmos.* New York: Basic Books, 2016.

———. *Does God Play Dice?: The New Mathematics of Chaos.* Oxford: Blackwell, 1990.

———. *In Pursuit of the Unknown: Seventeen Equations That Changed the World.* New York: Basic Books, 2012.

Stillwell, John. *Mathematics and Its History.* 3rd ed. New York: Springer, 2010.

Strogatz, Steven. *Nonlinear Dynamics and Chaos.* Reading, MA: Addison-Wesley, 1994.

———. *Sync: The Emerging Science of Spontaneous Order.* New York: Hyperion, 2003.

Sussman, Gerald Jay, and Jack Wisdom. "Chaotic Evolution of the Solar System." *Science* 257, no. 5066 (1992): 56–62.

Szpiro, George G. *Pricing the Future: Finance, Physics, and the Three-Hundred-Year Journey to the Black-Scholes Equation.* New York: Basic Books, 2011.

Tegmark, Max. *Our Mathematical Universe: My Quest for the Ultimate Nature of Reality.* New York: Knopf, 2014.

Thompson, Richard B. "Global Positioning System: The Mathematics of GPS Receivers." *Mathematics Magazine* 71, no. 4 (1998): 260–69. https://pdfs.semanticscholar.org/60d2/c444d44932e476b80a109d90ad03472d4d5d.pdf.

Turnbull, Herbert W., ed. *The Correspondence of Isaac Newton, Volume 2, 1676–1687.* Cambridge: Cambridge University Press, 1960.

Wardhaugh, Benjamin. "Musical Logarithms in the Seventeenth Century: Descartes, Mercator, Newton." *Historia Mathematica* 35 (2008): 19–36.

Wasserman, Steven A., and Nicholas R. Cozzarelli. "Biochemical Topology:

Applications to DNA Recombination and Replication." *Science* 232, no. 4753 (1986): 951–60.

Westfall, Richard S. *Never at Rest: A Biography of Isaac Newton.* Cambridge: Cambridge University Press, 1981.

Whiteside, Derek T., ed. *The Mathematical Papers of Isaac Newton, Volume 1.* Cambridge: Cambridge University Press, 1967.

———. *The Mathematical Papers of Isaac Newton, Volume 2.* Cambridge: Cambridge University Press, 1968.

Whiteside, Derek T. "The Mathematical Principles Underlying Newton's Principia Mathematica." *Journal for the History of Astronomy* 1, no. 2 (1970): 116–38. https://doi.org/10.1177/002182867000100203.

Wigner, Eugene P. "The Unreasonable Effectiveness of Mathematics in the Natural Sciences." *Communications on Pure and Applied Mathematics* 13 (1960): 1–14. https://www.dartmouth.edu/~matc/MathDrama/reading/Wigner.html.

Wisdom, Jack, Stanton J. Peale, and François Mignard. "The Chaotic Rotation of Hyperion." *Icarus* 58, no. 2 (1984): 137–52.

Wouk, Herman. *The Language God Talks: On Science and Religion.* Boston: Little, Brown, 2010.

Zachow, Stefan. "Computational Planning in Facial Surgery." *Facial Plastic Surgery* 31 (2015): 446–62.

Zachow, Stefan, Hans-Christian Hege, and Peter Deuflhard. "Computer-Assisted Planning in Cranio-Maxillofacial Surgery." *Journal of Computing and Information Technology* 14, no. 1 (2006): 53–64.

Zorin, Denis, and Peter Schröder. "Subdivision for Modeling and Animation." *SIGGRAPH 2000 Course Notes*, chapter 2 (2000). http://www.multires.caltech.edu/pubs/sig00notes.pdfNotes.

Index